Cascades
OF THE
BIG SIOUX RIVER

Cascades
OF THE
BIG SIOUX RIVER

A NATURAL & CULTURAL HISTORY

PETER CARRELS

Published by The History Press
An imprint of Arcadia Publishing
Charleston, SC
www.historypress.com

Copyright © 2025 by Peter Carrels
All rights reserved

Front cover, top left: A strategically situated observation deck at Falls Park reveals dramatic aspects of the cascades. *Courtesy of artist Stephen Randall. Top right*: Lone Rock formation within the cascades was toppled by the ferocious Big Sioux River flood of 1881. *Courtesy of Robert Kolbe Dakota Collection. Lower left*: Early Sioux Falls residents witnessed a wild, untouched environment at the falls of the Big Sioux River. *Courtesy of the* Sioux Falls Independent. *Lower right*: Waterfalls and exposed Sioux Quartzite bedrock create a stunning setting at Falls Park. *Courtesy of South Dakota Tourism.*
Back cover, top: Park designer John Royster emphasized sidewalks rather than roadways so Falls Park visitors could appreciate intimate views of the cascades and panoramic views of the sky and Sioux Falls. *Courtesy of South Dakota Tourism. Circular image*: The rugged, scenic falls of the Big Sioux River presented an unexpected landscape within a region dominated by lush, sprawling prairie. *Courtesy of Siouxland Heritage Museums.*

First published 2025

Manufactured in the United States

ISBN 9781467170055

Library of Congress Control Number: 2025936317

Notice: The information in this book is true and complete to the best of our knowledge. It is offered without guarantee on the part of the author or The History Press. The author and The History Press disclaim all liability in connection with the use of this book.

All rights reserved. No part of this book may be reproduced or transmitted in any form whatsoever without prior written permission from the publisher except in the case of brief quotations embodied in critical articles and reviews.

I am indebted to Betsy and Ed Marston who patiently, very patiently, encouraged me to understand the application and value of concise, impassioned and principled journalism. They gave me my first opportunity to write about natural resources, stewardship and environmental issues, and they published my pieces in the remarkable publication they led, High Country News. *I was one of many who benefited from their mentorship.*

CONTENTS

Preface 9
Acknowledgements 11
Introduction 13

1. Origins 17
2. Forebears 31
3. Sunset Land 49
4. Another Rebound 59
5. Industrialization 71
6. The Saga of Seney Island 99
7. Lily of the West 109
8. Hazel 119
9. Providence and Place 135

Bibliography 165
Index 169
About the Author..176

PREFACE

We never tire of waterfalls. Our ancestors experienced what we experience: the rhythmic melody of falling water that can purr or roar. The delightful sight of it. A hissing blur. Relentless, even when trickling.

For nearly forty years, starting in the late 1970s, I worked both as an activist and a journalist on challenging, sometimes contentious issues related to rivers and the environment. I was especially interested in the underappreciated James River in the eastern Dakotas and the brawny, over-manipulated Missouri River. I traveled across much of the western United States learning about, writing about and publicizing problems and wonders related to rivers and nature. I was exposed to topics ranging from industrial and sustainable agriculture to endangered and thriving species and wrecked, salvaged and saved ecosystems. At the end, much of my work contested atmospheric pollution and climate change. Rivers were helpful barometers.

In 2015, my wife and I moved from Aberdeen, South Dakota, our hometown, to Sioux Falls, South Dakota, so I could write and produce magazines at the University of South Dakota's Schools of Medicine and Health Sciences. Combative environmental politics had worn me down, and the nonconfrontational, collaborative atmosphere associated with public-spirited higher education and health care was a tonic. Not long after I started work at the university, the founder of a new nonprofit group based in Sioux Falls called Friends of the Big Sioux River invited me to join the organization's board of directors. I was grateful to work with the Friends group. My involvement with rivers would continue, including modest opportunities to make a small but satisfying difference.

Preface

Friends of the Big Sioux River was making friends along the river. Those first years of the organization's existence were spent establishing the group's reputation and connecting the community to the character and needs of the river. When I proposed a printed newsletter, the board quickly said yes. For seven years, I wrote and managed that biannual bulletin I first titled *The Advocate* and later renamed *The Otter*.

Writing and researching the newsletter provided an incentive to learn about the river. I interviewed engineers, geologists and other experts; examined studies and reports; and worked with archivists to find old photos and relevant communications. It was through newsletter-related inquiries that I became curious about the history of the river's cascades, the namesake of my new community.

I often visited Falls Park. And although it is situated near Sioux Falls' commercial district and a massive livestock slaughterhouse and meat factory, it was not an unreasonable place to commune with nature. At Falls Park, I was surrounded by happy families, out-of-town visitors, dog walkers, bicyclists and teenagers daring each other to edge close to rushing water. It was exhilarating to see so many people gazing at a river plunging over stacks of striking purple, reddish rock. I suspect that for many of them, it was a rare occasion to witness up close the evident and veiled beauty of a living river. Some of them, it was obvious, were encountering the cascades for the first time. I saw them stop in their tracks, caught off-guard by such a breathtaking and unexpected place in the heart of a busy city.

The human history of the cascades in Sioux Falls conforms to a pattern seen in communities across the country. Residents squandered a local river, came to their senses and worked together to recover what was finally understood to be a public resource. Although an initial emphasis of this rising environmental awareness was to honor a river's scenic attributes, in wiser communities a river's biological and ecological values were also acknowledged, addressed and protected.

Rivers are seldom harmed by accident and never healed by magic. Citizens are learning that the condition of a river can be determined by choices, and those choices are based on social priorities and ideals.

ACKNOWLEDGEMENTS

Special thanks to Jessie Nesseim at Siouxland Heritage Museums; Brett Kollars, assistant director of parks and recreation for the City of Sioux Falls; John Royster; Mike Cooper; Wayne Fanebust; Eric Dalseide; and Siouxland Libraries.

Interviews were conducted with Richard Burns, Mike Cooper, Tim Cowman, Jeff Danz, Travis Entenman, Wayne Fanebust, Adrien Hannus, Gary Hansen, Bill Hoskins, Jon Jacobson, Carter Johnson, Rick Knobe, Robert Kolbe, Brett Kollars, Ed Monson, Dave Munson, Barbara O'Connor, Sam Ogdie Jr., Jake Quasney, John Royster, Jeff Scherschligt and Craig Spencer.

Research assistance was supplied by Big Muddy Workshop, Burlington Northern Santa Fe Railway, Center for Western Studies at Augustana University (Elizabeth Cisar), Confluence (Jon Jacobson), Dubuque County Historical Society Museum, Excel Energy, Friends of the Big Sioux River, Good Earth State Park, Izaak Walton League (Sioux Falls Chapter), Levitt at the Falls, Minnesota Historical Society (Yves Hoppie), Edward Raventon, Eric Renshaw, Sioux Falls Parks and Recreation Department, Sioux Falls Public Works Department, Siouxland Heritage Museums and Irene Hall Museum Resource Center (Kevin Gansz, Bill Hoskins, Jessie Nesseim), Siouxland Libraries, South Dakota Geological Survey (Tim Cowman), South Dakota State Historical Society, State Historical Society of Iowa, University of Sioux Falls/Mears Library, University of South Dakota

Archives and Special Collections (Doris Peterson), the History Club (Sioux Falls) and Yankton Community Library.

The following people reviewed sections of the manuscript: Mike Cooper, Tim Cowman, Wayne Fanebust, Adrien Hannus, Gary Hansen, Carter Johnson, Mitch Kannenberg, Dave Munson and John Royster.

Please note that historians, researchers, explorers, residents and scientists use the words *falls* and *cascades* interchangeably. That practice is followed in this book.

INTRODUCTION

There are no unsacred places; there are only sacred places and desecrated places.
—Wendell Berry

Longtime Sioux Falls journalist Carson Walker liked to start his days with a run or a bike trek following the city's greenway trail system, and he often found himself at Falls Park as the sun rose and spotlit rushing water and quartzite rocks. The setting was stunning, said Walker, and as a writer and storyteller, he couldn't help but imagine the reactions of those who witnessed the same scenery during previous decades. "Were their impressions and sensations similar to mine?" he asked. "Probably not."

"Thirty-five years ago, in 1989, I arrived in Sioux Falls and first visited Falls Park," remembered Walker. The young reporter discovered a ragged, unkempt place. "In those days," he explained, "a visit to the falls might not be a pleasant experience. If you wanted to score drugs or hide from society, you would go to the park."

Over the years, park conditions changed for the better, said Walker. More recently, on a tranquil sunny morning, Walker sat on a slab of quartzite stone and studied the remains of the Queen Bee mill and other industrial reminders scattered across the park. "There were many different circumstances and purposes at the falls during the past couple centuries, and so many different emotions felt by those who came here during those many years," Walker reflected. "If only the rocks could talk."

INTRODUCTION

Sioux Falls, 1872. *From* The North and West Illustrated *by Chicago and North Western Railway Company (W.H. Stennett, 1872).*

Even a voiceless stone can tell stories. There are gouges and scratches caused by glaciers and abrupt edges sliced by quarrying cutters. Fortunately, there are human narrators to document the recent history of the cascades of the Big Sioux River.

Early visitors who stood beside the falls praised the site's unique beauty. Those intrepid explorers, scouts and settlers had crossed many miles of wild prairie before entering a geographically compact realm defined by waterfalls and exposed bedrock. Not only were the cascades of the Big Sioux River ruggedly majestic, they were also a regional anomaly, a geological formation out of sync with its surroundings. That added to the location's allure.

Written descriptions recorded in the years immediately before and after settlement of Sioux Falls reveal admiration and veneration toward the cascades. "[We came to a] great and picturesque fall of the river.…These falls present a remarkable feature of the river and country," wrote U.S. Army Captain James Allen during an 1844 reconnaissance of the region. A townbuilding company intent on earning a fortune at the cascades colorfully described the iconic landmark. The falls, exclaimed the Dakota Land Company in 1859, "is picturesque beyond description.…The skillful hand of nature seems to have excelled itself here."

As Sioux Falls' economic expectations fattened, the city's relationship to the cascades changed. Quartzite quarries disfigured the landscape. Industry bullied the river. And Sioux Falls did its damndest to ruin the entirety of the cascades area. The shoreline bordering the falls was sprinkled with water polluters, racket-makers and scruffy infrastructure. Dams and weirs in the channel blocked and redirected natural flows.

The United States and Sioux Falls were in no mood to consider conservation or preservation of natural resources. Everything was fair game; everything was considered raw material. Development and economic growth were based on the exploitation of flora, fauna, waterways and landscapes.

In 1908, the Sioux Falls *Argus Leader* newspaper celebrated the engineering prowess of a company opening a large hydropower facility alongside the falls. The engineers, reported the paper, had achieved a "victory" by successfully shackling the river and harnessing wild flows.

A new era brought new sensitivities. The conservation and environmental movement gifted society with broader perspectives and considerations. Community activists rose up to defend the cascades. Resurrection of the city's namesake following decades of abuse included the creation of Falls Park.

Carson Walker understood that the rocks at the cascades had witnessed a history worth telling. I'd like to think that if those rocks could talk, what's in this book is part of what they'd say.

Chapter 1

ORIGINS

Earth's geological diary has been written through millennia onto rocks, dirt, fossils and water spread across hillsides, canyons and plains. It is a journal comprised of broad, sweeping statements and intimate glimpses. There are entries about a single moon in the sky and tiny particles in the atmosphere, about jagged peaks, glaciated lowlands, vast lakes and modest ponds. There are passages describing pools of oil, grains of sand, rising magma and boulders half-buried in a meadow. All places on earth are a sum of processes, parts and time.

Understanding the creation and evolution of earth's surface is a vexing assignment. There were no human eyewitnesses to critical events that shaped this world. There are no firsthand accounts explaining the carving of a coulee or the filling of an ocean. Nearly all of earth's geological history occurred long before the earliest two-leggeds wandered land and seas.

Earth's exterior was fashioned over billions of years, an unfathomable span of time. Nonscientists may attempt to comprehend earth's contours and physical characteristics by referencing fleeting moments of high drama in the present tense. Perhaps there was a mudslide or an avalanche, a raging flood or a forest fire, a tornado or a typhoon. Harder to appreciate are the slow-paced consequences of colossal and far-reaching events such as a tectonic plate shifting or a glacier receding.

Trained geologists use knowledge and technology to interpret landscapes that appear as bewildering puzzles to the untaught eye. Geology is a relatively new discipline, gaining popularity between 1830 and 1840. A subset of geology is geomorphology, the study of landform origins and those processes that produce the world's relief features. James Edward (J.E.) Todd, an Ohio native, served as South Dakota's first state geologist. Todd was educated at Oberlin, Yale and Harvard before arriving in South Dakota in 1892 to teach geology and mineralogy at the state's newly minted university in Vermillion. The following year, the South Dakota legislature created a geological survey department to serve the state, and Todd was hired to direct that institution, a position he held until 1903.

Twenty-eight years before South Dakota organized its own geological survey, a national geologic agency had been assembled within the federal government. It was in the public's interest, declared the agency's creators, that mineral and other resources in the national domain be identified, located and evaluated. Economic prosperity, they emphasized, would benefit if citizens possessed a more comprehensive knowledge of the country's natural abundance. Decades before a conservation ethic was expressed, harvest and extraction were the only considerations by American civilization.

Agency scientists and surveyors studied stream depths, river basins and soil types. They suggested mining opportunities. They identified trees for logging and quick-falling waterways for hydropower. The intelligence and data generated by geological surveys motivated settlement and economic expansion.

In his initial report to the South Dakota legislature, in 1894, Todd noted the state's abundant natural resources and emphasized the whereabouts of "economic geology," describing mineral ores, building stones, artesian wells, lignite deposits and other bounty. Included in his analysis were glowing marketplace predictions regarding the commercial attributes of the pink and reddish rock visible at the cascades of the Big Sioux River. Todd also saluted the commercial potential of the river itself. "One of the finest opportunities for waterpower," he stated, "is at Sioux Falls."

Todd understood that every river grows flowing downhill. He knew that most river headwaters are situated in sparsely populated spaces, and the Big Sioux was no different, beginning far upriver from Sioux Falls as an outflow from a series of ordinary marshes and ponds in South Dakota's Roberts County. This is a sky-dominated locale, with few trees to block the view. A visitor walked the windswept terrain of the Big Sioux's birthplace and observed that early on, the young river is merely a thin, shallow crease crossing rich, fertile land. The Big Sioux courses southward, draining an interconnected hydrological territory, or watershed, measuring about nine thousand square miles and touching three states. Water feeding the

Big Sioux moves through hundreds of tributaries large and small before reaching the mainstem.

Comprising much of the Big Sioux's watershed is a flatiron-shaped, rolling plateau fringed in some places by abrupt slopes. Rising conspicuously between lowlands to the west and east, this highland was first called the Coteau des Prairies, or prairie hills, by early French fur traders and explorers. The coteau was formed and built over several million years, starting about three million years ago, when sedimentary debris from early continental glaciers was deposited on top of preexisting bedrock. Subsequent glaciers dropped more sediment, causing the plateau/coteau to grow and gain elevation. The surface of today's coteau is pocked with ponds, lakes, streams and one main river, the Big Sioux. The coteau rises near the North Dakota–South Dakota border and slopes southward. The lowest portion of the coteau, its southernmost edge, is located near Sioux Falls.

Among the world's landforms are a multitude of similarly featured expressions of geology. But there is only one prairie coteau.

The Big Sioux River longitudinally bisects the coteau, its channel tracking the tilt of this topographic formation. J.E. Todd described the river's elevation drop in his 1894 report. At its source, wrote Todd, the elevation of the Big Sioux is about 1,826 feet. Near the river's mouth, where it enters the Missouri River, the elevation is 1,098 feet, for a total drop of 728 feet and an elevation change averaging about 1.7 feet per mile.

Many rivers descend most dramatically in that portion of their basin nearest the headwaters. But the Big Sioux presents a different topography. In the stretch starting at modern-day Sioux Falls, roughly the lower third of the river's length, the channel loses the same amount of elevation that it loses above Sioux Falls. Much of that drop happens at the plunging, half-mile-long cascades of the river.

In its southernmost section, below the cascades, the Big Sioux runs beneath husky bluffs crowded with hardwood forests. The river's energy and capacity become stronger and deeper, and the channel becomes noticeably larger. In the earliest years of Euro-American settlement, business speculators were mistakenly confident that a commercial navigation industry could exist on this stretch of river.

Finally, 420 river miles from its source, water from the Big Sioux and its watershed vanish into the much larger Missouri River. At this confluence, the Big Sioux, during high flows, can muscle its way a long distance into the flowing Missouri before being swallowed up. The mouth of the Big Sioux occupies a geographically significant location: it is situated at the southeastern tip of South Dakota and is bordered by Iowa on the east and Nebraska to the south.

Although river flows animate the falls of the Big Sioux River, rock formations configure and define how water moves and tumbles through this distinct landscape. That stone, a bedrock called Sioux Quartzite, outcropped in several Sioux Falls locations, and the city's promoters and quarriers would eventually come to understand that they needed to contrast the commercial virtues of their local stone with those of granite in order to gain favor in highly competitive construction and building markets.

On a grassy crest above a steep ravine and a small tributary stream several miles from the cascades of the Big Sioux River, a chunk of unassuming fieldstone protruded from the soil. This cantaloupe-sized granite fragment colored a speckled gray had likely been pruned by a massive glacier from a stone formation many miles away, perhaps as far north as the Canadian Shield, one of the oldest visible rock formations on earth. That shard of stone had been carried a great distance by that growing sheet of ice and snow. Although the age of that small rock likely exceeds four billion years, it had arrived far more recently at its new home. Some twelve thousand years ago, that piece of granite was deposited onto or into earth's crust by a thawing glacier.

Earth's surface layer, the crust, is a sheeting of rock and other materials covering the outermost portion of the planet. With an average thickness of ten miles, the crust composes less than 1 percent of earth's mass, though it does serve as the floor for all oceans, except the deepest Pacific. Crumbled stone and varying sediments and soil types also compose a portion of the crust.

The crust is underlain by the mantle, a thicker band of rock and gas also encasing the globe. Containing about 80 percent of earth's mass, the mantle is a relatively stable formation, while the crust subtly shifts and floats. Beneath the mantle, at the center of the globe, is the mysterious core, composed of a hardened inner sphere and an outer layer of searing, low-viscosity fluid. Although temperatures within the inner core are exceedingly high, pressures are intense, preventing rocks in the hottest portion of the planet, the very center, from dissolving and liquefying.

Granite in its most rudimentary forms was sired by molten or partially molten rock called magma. Creation of this foundational material typically occurred in earth's mantle or in chambers beneath volcanoes, where the temperature of liquid rock ranged from 1,400 to 2,200 degrees Fahrenheit. The appearance and structure of granite is an outcome of how magma reached earth's surface.

After magma emerges from inner earth, it transforms to lava. Magma slowly ascending to earth's surface through vents and fissures bubbled through the crust and cooled slowly. This type of lava, nonvolcanic, was eventually transformed into a type of granite that is coarse and may contain mixed markings or speckles. Volcanic magma, thrust suddenly into the air, undergoes a faster cooling process, creating granite that is glassy or smooth.

The quartzite shaping the cascades is younger than granite by at least a couple billion years and has been fashioned by different forces.

An enormous inland sea pooled where eastern South Dakota and other parts of the Northern Plains are now. During its lengthy lifespan, that sprawling, fluctuating body of water generated shifting shorelines, and quartz sands carried by waterways or blown by winds accrued at the ever-changing shallow edges of that sea.

Over millions of years, those sand accretions were transported by rivers to downstream locations. Sands at the base of the concentrated deposits were compacted by overlying sediment. That process, the downward force of heavy overburden, squeezed excess liquid from the sand, like pressing down on a water-soaked sponge. The compressed sand grains were then locked together by silica, and sandstone was formed. Five million years later, after additional weighty compression, the sandstone was squashed into Sioux Quartzite. Tinted red, pink or purple by iron oxide, this quartzite possesses a texture that is tightly interlocked. In other words, it is a rigid, exceptionally hardened stone.

Geologists typically refer to Sioux Quartzite as a sedimentary stone, although they acknowledge that this quartzite is most accurately described as a hybrid, blending characteristics of both sedimentary and metamorphic rocks. Metamorphism involves the influences of heat and pressure, causing rock or rock elements to recrystallize, resulting in an altered texture or mineral composition. Sedimentary rocks formed after mineral or organic particles on or near earth's surface were dispersed and deposited by wind or water. This type of stone often appears as layers or strata when mined below ground or when visible above ground. Granite is an igneous rock, a type of stone formed following the cooling of magma or lava.

Sioux Quartzite, mostly buried beneath glacially deposited soils and stony sediment, is classified by science as a bedrock. This handsome stone was revealed to sunlight and weather by stream action and glaciers at select locations throughout its range, which includes southwest Minnesota, northwest Iowa, southeast South Dakota and northeast Nebraska. Stone similar to Sioux Quartzite is found at few places on earth.

Tim Cowman, who heads South Dakota's Geological Survey at the University of South Dakota, explained the stone's physical significance. "Created during Precambrian time," said Cowman, "this quartzite forms a deep layer of rock that is part of the foundation on which the rest of the region's geologic history is built."

In 1867, a geologist named C.A. White visited newly settled Sioux Falls, then a quiet hamlet in Dakota Territory. A small military post and a handful of settlers huddled there. White, a Massachusetts native, studied geology at the University of Michigan and medicine at Rush Medical College. He moved his family to Iowa City, Iowa, to open a medical practice but soon realized his real passion was geology and natural history, so he switched careers and was appointed to direct Iowa's geological survey. White would become one of the nation's most respected scientists, publishing over two hundred papers and books and claiming affiliations with leading science institutions, including the U.S. Geological Survey and the National Museum (the original name for the Smithsonian Institution).

An early expedition during White's decorated career was his visit to the valley of the Big Sioux River. White and five companions departed Sioux City, Iowa, and headed north, following the east side (the Iowa side) of the Big Sioux. White soon noted the appearance of red boulders "embedded in the deep, rich soil." Farther north, he encountered "ledges of the red quartzite," and as the group neared the settlement called Sioux Falls, they found quartzite exposed at regular intervals along the valley.

In Sioux Falls, the Iowa scientist studied the cascades of the river and the prominent quartzite formation observable there. In an article published by *The American Naturalist*, he described the setting: "We find a magnificent exposure of the same rock extending across the river and causing a series of falls of sixty feet in aggregate height, within the distance of half a mile, which for romantic beauty are seldom surpassed."

White wrote descriptively and knowledgeably about the stone.

> *This quartzite is of a nearly uniform brick-red color, intensely hard, quite regularly bedded, the bedding surfaces sometimes showing ripple markings as distinct as any to be seen upon the sea-shore of the present day, and which were made in the same manner untold ages ago, when this hard rock was a mass of incoherent sand, the grains of which are even now distinctly visible. In a few localities it presents the characters of conglomerate, the pebbles being as clearly silicious as the grains of sand.*

A Natural & Cultural History

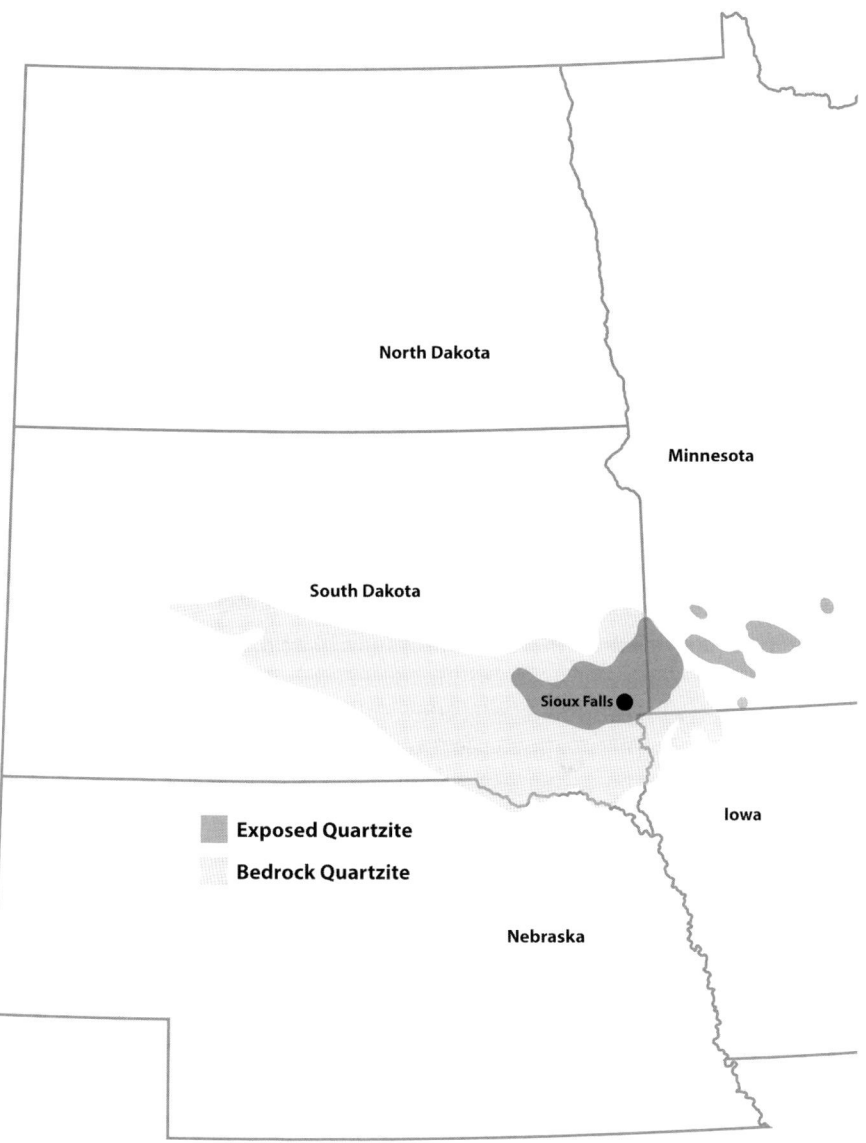

Range of Sioux Quartzite. This bedrock, also known as jasper stone, outcrops in few locations, including at the cascades of the Big Sioux River. *Map by Eric Dalseide.*

White spent several days at the falls before continuing northward to study the soft quartzite rock at the Pipestone quarry, the renowned Native American stone excavation site. He is now recognized as the first person to use the term *Sioux Quartzite*.

Eighty-two years after White's scientific mission to Sioux Falls, another geologist, Brewster Baldwin, released a report about Sioux Quartzite. Baldwin, employed by South Dakota's Geological Survey, was precise in his description of the stone. "The color of the quartzite," he wrote, "is commonly red or pink, but it can vary over a considerable range. Some exposures are nearly gray or white, with only a faint pinkish cast, and otherwise nearly black. A dark reddish purple is common, and in some places the rock has an orange tint." Baldwin also explained the variety of colors: "The color of the quartzite is due to the presence of thin films of iron oxides coating the grains of quartz."

J.E. Todd described a Sioux Quartzite quarrying attribute: "Although very hard," he wrote, "natural jointing of the rock makes it possible to excavate it with comparative ease."

The time of continental glaciers, the Ice Ages, began about two million years ago, during a period known as the Pleistocene epoch. Glaciation impacted present-day Canada and extended as far south in the United States as Kansas. Continental glaciers, as distinguished from mountainous, alpine glaciers, were enormous, expansive sheets of moving ice fed by snowfall. These ice sheets were born in northern polar regions and slowly advanced southward. Protracted periods of cold temperatures created the conditions necessary to produce continental glaciers. Generally speaking, growing glaciers behaved like rivers. They flowed downhill and followed the path of least resistance. Glaciers might be thin (hundreds of feet) or thick (more than one mile). Some glaciers existed for twenty thousand years, others for just several thousand.

Ice sheets didn't happen by accident. A slight shift in earth's orientation to the sun could bring about a gradual climatic transition. Glaciers didn't arrive in steady, regularly sequenced succession. Cold spells were intermingled with warmer periods that caused an existing, moving glacier to slow, stall or regress northward. Many thousands of years could separate glacial events. A glacier might pause for a century or more due to warming temperatures and then resume its journey as colder weather again compelled the ice sheet to expand and push farther south. Each individual glacier intermittently receded, following a time frame dependent on climate, before finally disappearing except for frozen sections nearest the polar ice cap.

Spreading like wax from a burning candle, an advancing glacier smothered, crushed and bulldozed everything in its path. One common feature of all glaciers was the presence of meltwater at the base of the ice mass. This allowed a glacier to slide more easily across earth's surface. The pace and impact of a growing glacier was slow and methodical, not furious or frenetic. An infant might out-crawl a glacier, but it would require countless crawling generations to safely stay ahead of the moving ice.

The melting and waning processes of glaciation were also impactful to the land. Starts and stops interrupted glacial retreat, causing features like ice tunnels and moraines to build up where a stationary glacier warmed for many months or years. A retreating, thawing ice sheet produced massive volumes of meltwater and deposited immense quantities of clay, silt, sand, pebbles, cobbles and boulders, known as glacial till. This till had been harvested from the land and crushed and mixed within the churning mass of ice as a glacier pushed powerfully southward. Soils and veins of gravel and sand were created—as well as new layers of earth's crust. Meltwater surged downhill, establishing incised trenches that became rivers and streams.

Geomorphologists explain that the series of glaciers influencing the contours and character of today's landscapes began roughly sixty-five thousand years ago during the so-called Wisconsin glacial era. Over fifty thousand years, four distinct ice sheets moved south at varying intervals onto what is now eastern South Dakota. Each succeeding sheet of ice overrode and erased some surface features wrought by predecessors.

All Wisconsin-era glaciers pushed down from the far north until they encountered the raised topography of the prairie coteau. Unable to overtop this upland, a glacier would cleave into two separate ice sheets, or lobes, starting at the prow of the coteau: imagine the top of an inverted *U*. Those lobes, called the James lobe and the Des Moines lobe, veered west and east, respectively, on either side of the coteau, forming what would become today's James River and Minnesota River lowlands. A portion of the eastern ice sheet also established the Des Moines River valley.

When temperatures warmed and the glacial lobes began to thaw, meltwater traveled downhill, filling and overflowing low spots and carving what would become the circuitous routes of the modern-day James, Minnesota and Des Moines Rivers. The earliest version of the Big Sioux trench was created forty thousand years ago by meltwater moving southward, starting at a point where the two lobes split from the main ice sheet. This is why geologists call the Big Sioux an interlobate waterway, meaning the river lies between lobes. While meltwater from the two lobes

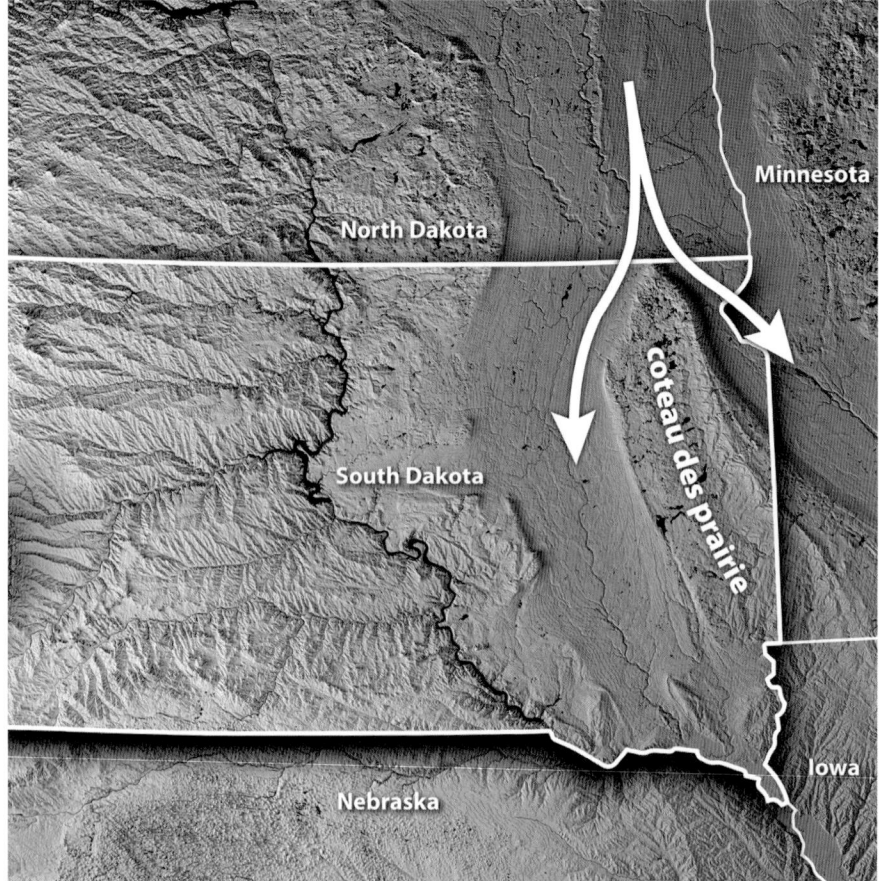

Above: Continental glaciers spreading south encountered the elevated prairie coteau and were deflected east and west. *Map by Eric Dalseide.*

Opposite: The Big Sioux River drains a watershed that mostly corresponds to the prairie coteau. The river's cascades are found in the heart of Sioux Falls. *Map by Eric Dalseide.*

surged through valleys and lowlands flanking either side of the coteau, the growing Big Sioux River also carried glacial thaw and sliced a channel through the plateau in the heart of this highland.

The glacier known as the late Wisconsin glacier, the most recent continental ice sheet, behaved like previous ice sheets. It advanced southward until it was blocked by the elevated northern extremity of the coteau and then split into two lobes. Once again, the lobes veered into the lowlands bordering the western and eastern margins of the coteau and each grew and progressed southward.

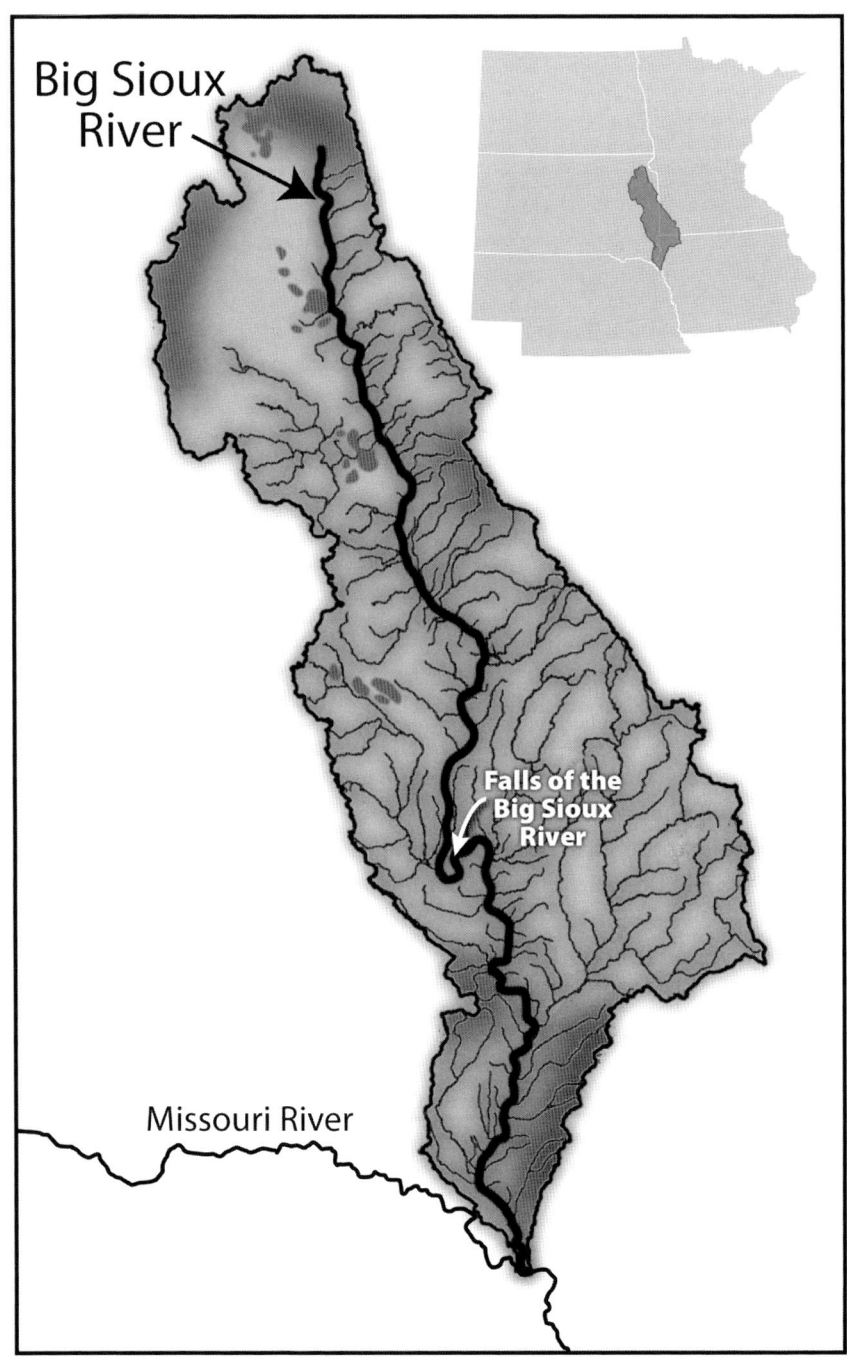

Immense, powerful glacial runoff peeled away a veneer of silt and soil to expose Sioux Quartzite bedrock at the cascades of the Big Sioux River. *Courtesy of South Dakota Tourism.*

At the southern and lowest portion of the coteau, where present-day Sioux Falls now sits, the James lobe (the western lobe) was finally able to spread east. Geologists estimate that the James lobe occupied this area for five thousand years, starting about thirteen thousand years ago. The features and contours of the terrain that emerged from beneath that ice sheet became the freshest and most visible glacial influences on today's Siouxland landscapes.

For several thousand years, this portion of the eastward-expanding James ice sheet blocked an early version of the Big Sioux River, forcing the river to swerve north and east for a short distance. That pronounced loop, where the river channel was redirected, remains evident, and it is there, on that stretch of channel, that the river's cascades formed. J.E. Todd identified that loop, asserting, "The peculiar bend on the Big Sioux River by which it was thrown over the quartzite at the Falls is to be referred to the presence of the ice sheet immediately south at this stage."

Richard Flint further explained the course of the river, describing the falls as occurring "in a part of an intricate S-bend in the Big Sioux trench," and he noted that another South Dakota geologist, Edgar Rothrock, had correctly interpreted how the river, valley and falls were formed in the Sioux Falls area. Rothrock, reported Flint, postulated that a large volume of glacial meltwater flowing southeasterly within the Big Sioux valley was blocked by an ice sheet that existed in today's southern Sioux Falls. The obstructed and diverted water, wrote Flint, "found an outlet northeastward down a short tributary to another parallel valley, and thence detoured the margin of the ice."

This powerful torrent, carrying an immense volume of thawed ice and snow, peeled away a veneer of glacial till, exposing quartzite bedrock. It was probable, Rothrock and Flint concluded, that these flows forcefully battered that mass of revealed rock, causing cracks and fractures and scattering bulky fragments across the channel and shore. When the pace of epic erosion subsided, the ancient Big Sioux River presented itself through half a mile of spectacular turbulence, first descending a broad stretch of symmetrical, stony stairsteps before plunging over tall ledges and rushing through narrow chutes, creating a series of steep, dramatic cataracts. The place had been transformed by extraordinary natural forces into a singularly unique and breathtaking landscape.

Chapter 2
FOREBEARS

The exact date a city council adopted plans to develop a new street or sewage treatment plant may be ascertained, but there is no way to know for sure when the earliest version of the flowing Big Sioux River etched a trench through an uplift on an ancient plain. Nor is there a reliable chronicle revealing when humans first stumbled upon the cascades of the Big Sioux River.

According to ecologist Carter Johnson, before Euro-Americans arrived equipped with their era's industrial technologies, the postglacial environment situated at the falls of the Big Sioux River hosted a variety of plant and animal species not found together elsewhere. The area was, Johnson explained, "an ecological gem, rich in biodiversity."

In addition to exposed quartzite bedrock, there was a living, healthy river that enabled assorted aquatic species to survive drought and prolonged cold by escaping to deep pools. "The aeration at parts of the falls," said Johnson, "would have improved oxygen levels for fish compared to the flatwater conditions of a seasonally dry river."

There were also shady riparian woods and endless prairies on the uplands. At the falls, the midgrass province of the plains rubbed up against denser tallgrass prairie existing at the western margin of its range. At this crossroads, flora and fauna from different regions intermingled.

After the final continental glacier of the Pleistocene epoch receded north, the region's first human visitors wandered from the west and northwest into the basin drained by the Big Sioux River. Today's topography remains sliced

Before Sioux Falls was settled, the cascades area was an ecological gem, providing prairie, woods and deep riparian pools. *Courtesy of Robert Kolbe Dakota Collection.*

by ravines and valleys, dented by lakes and topped by hills covered with fertile soils. There have been dramatic changes to ecosystems, no doubt, especially from impacts meted out by the most recent human residents, but the physical structure of the rural countryside bears a perceptible resemblance to the prehistoric terrain of this place.

The climate of the area now occupied by Sioux Falls was gaining warmth and drying out when the last ice sheet evaporated into history. A frozen land defrosted, allowing plants to prosper on glacial till. In certain places, spruce forests thrived. Migrating from the southwest, perennial prairie plants, including blazing star, prairie smoke, butterfly weed and smooth blue

The rugged, rocky cascades were an unlikely, unexpected geographical feature in a region dominated by smooth terrain and grasslands. *Courtesy of Robert Kolbe Dakota Collection.*

beard tongue, took root and claimed dominion. The age of megafauna, a time when there was a preponderance of large animals, would soon end in widespread extinction explained by differing theories.

Giant beaver, big as bears, disappeared. Mammoths and mastodon vanished, too. Species of bison, deer and antelope, larger than today's varieties, also perished. Imposing predators such as saber-toothed cats, dire wolves and the short-faced bear, deprived of food, slowly ceased to exist. One extinction theory centered on the rising success of prehistoric hunters armed with improving weapons and strategies. Another explanation identified the culprit as a warming climate that changed environmental and habitat conditions.

What was the reaction by those humans who first stepped onto the quartzite shoreline alongside the cascades of the yet unnamed Big Sioux River? Churning water and reddish rock would have been visible from afar as their band descended a nearby hillside. On the lookout for vicious carnivores, such as gray wolves, cougars or grizzly bears, that pioneering party of perhaps ten or fifteen people paused before proceeding. In greater abundance were many of the same flora and fauna witnessed today. There was birdsong. Snakes slithered underfoot. A thirteen-lined ground squirrel scurried into a den. As they crossed the floodplain, the sound of forceful water became noisier, rowdier. This place announced itself. Footing grew trickier on angular heaps and smooth stacks of stone. And then these traveling companions crept to the river's edge.

After hiking many weeks across a vast expanse of grasslands, they'd stumbled onto a welcome, wondrous sight. Close at hand were ribbons of water wildly diving over rocky cliffs and charging through chasms and chutes. The type of stone causing the river to fall was different than anything they'd seen before.

Perhaps these onlookers had followed the river valley from some distant locality, wondering where the waterway would lead. Now they crouched on the stony shore, weapons held loose and easy, captivated by the setting. They bathed in backwaters and camped beneath a stand of trees on level ground. Guarded by a watchman and a flickering fire, they likely slept peacefully that night, calmed by the hum of an animated river. If they followed custom, they would stay for a few sunsets before moving on.

It is likely these people were Paleo-Indians, descendants of the Clovis culture who had walked the so-called land bridge from what is now Siberia across the Bering Sea to enter North America. These people were light travelers and frequently on the move. The land around them was unpeopled. Encountering another band of their kind was rare.

The Clovis culture caused its own demise as different groups dispersed after crossing the land bridge, spreading quickly throughout what is now western Canada and the United States. The distinct environmental characteristics of their new homes dictated specialized behaviors and survival tactics. Those of this culture who eventually found themselves residing in the desert learned to live differently than those who roamed the mountains or the plains. People of the Clovis culture became people of many cultures. The progeny of those humans who first stood beside the falls would soon understand that widely roaming a wild land in small bands was not a sustainable formula for long-term survival.

On rounded summits of rolling bluffs near the falls of the Big Sioux River are burial mounds of a more recent people who valued nature and honored their dead. At least five such mounds have been identified within Sioux Falls, and they are visible in what is now Sioux Falls' first park, called Sherman Park.

The people who built the burial mounds lived year-round in a small village located near the Big Sioux River upriver from the falls, in the vicinity of modern-day Minnehaha Country Club and the Great Plains Zoo. They were Indians of the so-called Late Woodland culture, sustained by both agriculture and hunting and among the first humans to use bows and arrows and grow maize. Archaeologists excavated one of the burial mounds and found interment remains 1,600 years old.

A more substantial and more recent Native American community was established along the Big Sioux River a short distance downriver from the falls. This once-thriving village was called Nixe (now known as Blood Run) and is part of modern-day Good Earth State Park. The site was desirable because of its proximity to water, timber, prairie and game. Another appealing aspect was its nearness to the revered Pipestone quarry, some sixty miles north. Ceremonial objects hand carved from the soft quartzite found there served important religious uses, and pieces of raw stone as well as handcrafted finished items were highly desired by traders.

Scholars disagree over various aspects of the Blood Run settlement, though it appears this community straddling both sides of the Big Sioux River originated in the 1500s. Tribes associated with the Oneota culture, including Ioway, Otoe, Omaha and Ponca people, occupied the site continuously over its three-century lifespan. Eventually, the Omaha tribe would come to dominate the village. The settlement's population may have reached hundreds of residents, though during trading fairs, several thousand Indians may have camped there.

Blood Run residents lived in bark-covered lodges shaped like modern-day Quonset huts. Their dead were placed in burial mounds constructed of soil and gravel, similar to what was used at the Sherman Park location. These villagers maintained productive gardens and welcomed French fur traders beginning in the 1600s. Beaver became a target for Omaha hunters and trappers and a valuable trading currency with the French. The village complex at Blood Run was abandoned by 1719 because of Sioux aggressions.

Historian and author Edward Raventon explored the relationship between the Omaha people and the falls. His inquiries revealed no information about specific connections.

Adrien Hannus, a longtime South Dakota archaeologist, had a plausible explanation. "The falls," said Hannus, "had to have been one of the most important markers for early people traveling to or within the region." But the lack of reviewable records sourced by the Omaha, Sioux or other tribes honoring or describing what surely must have been a significant place for Native Americans, he explained, was likely caused by the culture's oral process of recounting and honoring history. "The rituals and traditional belief systems of the Omaha and other native tribes don't leave physical evidence," Hannus said.

Scientific investigations to track or speculate about the relationship of Native people to the falls were also challenged by difficulties performing archaeological examinations there. Industrial and recreational uses pursued by non-Native people have stripped away artifacts that might have been recovered and analyzed. "We never had opportunities to do examinations at the falls," reported Hannus. "There were too many major alterations to the land around the falls."

George Kingsbury noted the relationship of the Sioux people to the falls in his book *History of Dakota Territory*. "The site of Sioux Falls," Kingsbury explained, "was a favorite resort of the Sioux Indians from the earliest period of exploration of the country [by Europeans]."

Another account was written by a Pennsylvanian named Amos Gottschall who visited Sioux camps during the 1870s. Gottschall described the cascades as an important ceremonial, meeting and camping place for Indian people. "The falls were a favorite [place] of the Indians from time immemorial," he reported.

There is no writer more closely associated with Sioux Falls than Frederick Manfred. Born in 1912, forty miles east of Sioux Falls, Manfred taught writing at Augustana College and the University of South Dakota and authored two dozen books. His publications were mostly frontier-era novels focused on the West and Upper Midwest. Before he died in 1994, he was credited with christening the four-state region surrounding Sioux Falls "Siouxland," a nickname that remains relevant.

In his 1959 novel *Conquering Horse*, Manfred described a band of Yankton Sioux Indians camped near the cascades of the Big Sioux River. These people, said Manfred, believed that a god lived behind the rugged falls, obscured by thundering water.

Esteemed naturalist Ernest Thompson Seton estimated that as many as four hundred million beavers populated North America when Europeans began establishing a foothold on the continent. Beaver numbers plummeted as the fur trade marched from east to west. When trappers and traders depleted beavers in one area, they simply pushed on, finding untapped territory for ongoing exploitation. The first Europeans in America had left behind an over-trapped, mostly beaver-less Europe. Fur merchants had done there what they were destined to do in the New World. A lucrative market for an international industry existed primarily because hats made from beaver were in vogue.

Pierre-Charles Le Sueur, a French fur dealer and explorer, reached the northern region of the Mississippi River watershed in 1683. He was only twenty-six years old when he joined other fur traders working the western wilderness, and he may have been the first white man to set foot in South Dakota and to explore the Big Sioux River. Le Sueur established friendly business relations with area tribes, eventually expanding his network of trading partners to include Omaha Indians living at Blood Run village.

Le Sueur paid close attention as fellow Frenchman Robert Cavelier, Sieur de La Salle, successfully ascended the Mississippi River by boat, traveling from the Gulf of Mexico to the river's headwaters. At the time, connecting French fur traders and trappers in the upper Mississippi River watershed with wholesalers and retailers in Europe required goods to pass through Canada. Le Sueur wanted to avoid Canada so he could escape Canadian taxation.

Le Sueur's alternative passage had him shipping his wares south on the Mississippi to oceangoing vessels waiting at the mouth of the great river. Apparently, Le Sueur used two different routes to reach the Mississippi. One may have begun at the Blood Run village, where bundles of fur were moved south by voyageurs floating small boats down the Big Sioux and Missouri Rivers to St. Louis. Larger vessels then carried the furs down the Mississippi to New Orleans.

A second route included an overland leg and also began at Blood Run. Traveling easterly on a faint trail across three hundred miles of prairie, Le Sueur's employees hauled their cargo by cart to Prairie du Chien, located near the confluence of the Wisconsin and Mississippi Rivers in what is now southwest Wisconsin. At Prairie du Chien, pelts were transferred to boats for the trip down the Mississippi.

Maps available to frontiersmen roaming the Upper Mississippi River region were nonexistent or rudimentary. Cartographers hadn't yet applied their science and skill to this part of the continent. Mapmakers needed to assess a landscape, or they needed a reliable eyewitness. And that is what led Le Sueur in 1701 to a French cartographer's studio in Paris. Guillaume Delisle, a rising star in his field, wanted Le Sueur to tell him about the strange, faraway land Le Sueur had come to know. Seven decades before the Revolutionary War, Le Sueur had successfully grown a tricky business in the uncharted heart of the continent. Few knew this place like Le Sueur. Delisle also sought input from other explorers and observers.

The maps born from the Delisle-LeSeur collaboration, first published in 1702, and hand drawn by Delisle, included information never before circulated through any network of publishing and cartography. It is possible these were the first maps to depict the western portion of the upper Mississippi watershed, including the Big Sioux River and unnamed lands in what is now eastern South Dakota.

Delisle, born in Paris in 1675, was mentored by his father, a geographer and historian who trained French nobility. Delisle's maps emphasized clarity and integrity, traits that were appreciated in an increasingly competitive profession. Though he drew his maps in Europe, Delisle was meticulous regarding his sources, and he would later gain fame as the first to craft an accurate map of Canada. His 1718 map of Louisiana Territory is considered one of the important cartographic accomplishments in the annals of North America. He made history when he established Texas as a place name.

Of special note on the Delisle-Le Sueur maps was the depicted location of the Blood Run village on the Big Sioux River, which was titled Maha, named after the Omaha tribe. Some French traders referred to the Big Sioux as the River of the Mahas. The trail connecting Maha to the Mississippi River was also represented. Delisle and Le Sueur labeled that path Chenandes des Voyageurs, or Track of the Voyagers. It is likely Le Sueur visited Maha more than once and he may have also followed the Chenandes trail. That this primitive road was represented on their maps revealed how important it was to Le Sueur. Curiously, the maps did not identify the falls of the Big Sioux River.

Nine decades after the Delisle-Le Sueur maps were published, a more detailed map showing the middle Missouri River region, including the lower Big Sioux River, was published by a mapmaker named John Evans. Evans, a

Welshman, conducted topographical calculations during several exploratory expeditions sponsored by the Spanish government, and he later assembled a detailed map of the middle reach of the Missouri River. What made his map so valuable was that the mouth of each river or stream entering that reach of the Missouri was identified. The Big Sioux River was designated "R de Seaux," or River of the Sioux. Evans did not travel upriver on the Big Sioux, and the river's cascades were not noted.

Perhaps the most consequential exploration in our nation's history was the Lewis and Clark Expedition from St. Louis to the Pacific Ocean and back. During that extraordinary undertaking, a troop of forty-five explorers, soldiers and boatmen completed an arduous and risky crossing, much of it by river, through what was then a mostly undocumented territory.

President Thomas Jefferson's 1803 purchase of the 828,000-square-mile Louisiana Territory from France instantly doubled the landholdings of the United States. A small section within that vast acquisition was the entirety of the Big Sioux River watershed and its falls.

Jefferson wanted to document the economic opportunities afforded by the nation's newest asset so he sent Lewis and Clark to identify and examine landscapes, geology, minerals, flora, fauna (including beaver) and other resources and to create detailed maps of their entire journey. Jefferson also wanted to know if there was an all-water route to the Pacific Ocean. It was vital, instructed Jefferson, that members of the expedition befriend and appraise the Native people they encountered.

When Lewis and Clark traveled upriver on the Missouri River starting at St. Louis in the spring of 1804, they possessed whatever available reconnaissance existed about Louisiana—and there wasn't much. Jefferson did provide them with a copy of John Evans's map. Enormous sections of their journey would be unmapped or poorly mapped and vaguely understood. But on that long stretch of the Missouri River flowing through much of today's North and South Dakota, Lewis and Clark continuously referred to the map prepared by Mr. Evans.

William Clark's stunningly detailed maps of the entire expedition route, completed four years after the Louisiana exploration concluded, identified the mouth and lower channel of the Big Sioux River. Clark included the river's name: Grand River de Sioux. An expedition journal stated that this river was "navagable to the falls 70 or 80 Leggues." The distance was inaccurate, but this may have been the first time a "falls" on the Big Sioux River had been cited in a written document or on a map. Although no one from the expedition scouted sufficiently upstream on the Big Sioux to view

> ## Louisiana Territory and an Evolving Nation
>
> The birth date of Louisiana Territory was April 9, 1682, when French explorer Robert Cavelier, Sieur (Lord) de La Salle, erected a Christian cross near the mouth of the Mississippi River. At a formal celebration, La Salle asserted that France was taking possession of the entire Mississippi River basin and that this claim was sanctioned by the grace of God. To honor sitting French King Louis XIV, La Salle named the land Louisiana.
>
> The French were initially intrigued about this remote part of the New World, but as they learned more about the region's endless plains, their interest waned. French scientist Georges-Louis Leclerc de Buffon described Louisiana as a primitive place "fit only for degenerate life-forms."
>
> After purchasing Louisiana from France, the American government designated two separate entities within the newly acquired land. Orleans Territory was created in the southernmost area, and the remainder was titled District of Louisiana. Lewis and Clark hadn't yet completed their important exploration, and the political boundaries of the territory they passed through had already been restructured.
>
> In 1805, the District of Louisiana was organized as Louisiana Territory. Seven years later, much of Orleans Territory was admitted as the eighteenth state of the United States, Louisiana, and what was left of Louisiana Territory was added to Missouri Territory.

the cascades, it's likely Lewis and Clark learned of them during consultations with a Sioux interpreter.

Clark's maps and the expedition's journals noted the presence of enormous populations of wildlife along the Missouri River and its tributaries. The news spread quickly. Opportunistic hunters, trappers and traders moved into remote reaches of the territory. Traffic on the Missouri River increased. Overland ventures into the region escalated. Artists and ornithologists eagerly journeyed to this paradise that some called the Serengeti of North America. The expedition led by Lewis and Clark unlocked Louisiana and the Pacific Northwest to American businesses, town builders and settlers. Young men in the nation's older states looked westward and yearned for adventure and improved economic prospects. One such man was Philander Prescott.

Prescott, born in 1801, and his five siblings had been orphaned and left to fend for themselves in their hometown of Phelps, a rural village in western New York. Prescott's older brother, Zachariah, left the family for employment as a clerk at a sutler's store in Detroit, Michigan, selling wares to soldiers. Curious about the American wilderness, teenaged Philander was thrilled when his older brother offered him a job. At the time, Detroit was a newly incorporated city on the eastern edge of the western frontier. The isolated outpost had, not long before, served as one of England's four major North American fur trading posts.

Prescott packed his meager belongings into a single bag and began the hike to Buffalo, New York, where he boarded a steamship for the rest of the trip. By the time Prescott arrived in Detroit, during the summer of 1819, the U.S. Army had already begun repositioning troops from there to stations along the Mississippi River, reflecting a frontier shifting westward. Prescott also journeyed west and secured work at Camp New Hope, situated near Fort Snelling at the confluence of the Minnesota and Mississippi Rivers in modern-day Minnesota. Prescott found himself on the harsh margin fringing America's newest frontier, and although living conditions were raw and rough, he was delighted at his prospects. He was determined to try his hand at the fur trade.

Over the next decade, Prescott bushwhacked and paddled back and forth across the Minnesota wilderness in all seasons, learning the skills of an explorer and leading others on those same missions. During an 1824 fur trading foray into what is now central Minnesota, Prescott and two other men pushed into Indian country. The trio was suddenly surrounded by a large group of Sisseton and Wahpeton Sioux. The youngest warriors wanted to kill the three interlopers, suspecting they were spies working with the Ojibway, fierce enemies of the Sioux. Prescott assured the Indians he only wanted to trade with them. He explained that he and his associates were lost. It seems Prescott effectively made his case. "The older men interfered and stopped the young from killing us," recalled Prescott. It didn't hurt Prescott's cause that he was married to a Sioux woman named Na-he-no-Wenah (Spirit of the Moon), the daughter of a subchief, a fact confirmed by one of Prescott's captors. After being held for three days, Prescott and his companions were finally released. Shaken, the men shouldered their packs and trudged homeward through a snowstorm.

Prescott and others in the fur trade endured brutal weather, hunger, exhaustion, injuries, sickness, fleas, mosquitoes, sleeping cold in wet clothes and learning to ignore the foul smell that was one's own aroma. They learned how to patch a canoe, read the sky to anticipate weather and pack for the trail. Prescott and his fellow adventurers could build a wooden hut in short order and survive in it through a merciless winter.

In the frontier fur business, trappers, traders and voyageurs moved between remote and isolated camps and outposts, often occupying sites on lands claimed by Native American tribes. Because fur traders furnished supplies of value to Native Americans, non-Natives involved in the fur business were tolerated in Indian country. By 1826, Prescott was one of only ten licensed fur traders working with the Sioux Indians.

Two principal fur trading companies operated in Minnesota: the Columbia Fur Company, with a dozen posts in the area, and the American Fur Company, with twice that many. New companies sprang up as well. Prescott and his brother Zachariah were original members of the Columbia syndicate.

Nothing about the fur trade was guaranteed. There might be poor management, labor shortages, supply scarcities, dwindling animal populations or troubles with Native Americans. There were dishonest partners and cutthroat competitors. The owners of the Columbia company eventually tumbled into impossible indebtedness and sold their inventory and supplies to the American Fur Company. Prescott and others were suddenly unemployed, a situation he described as, "[We] were left to grub for our lives."

Desperate to find work, Prescott left his family at Fort Snelling and traveled by steamboat to St. Louis, where his brother was working as a bookkeeper. To gain passage, Prescott served as a clerk and cargo watchman, and when the boat docked in St. Louis, the vessel's captain kept him on as a freighter, loading and unloading the steamboat through the winter as they boated trade routes in the south. It was a trying time, away from his family, and Prescott finally boarded a boat bound for home. He found his wife and children in the town of Mendota, near the fort, living with his Indian in-laws. Na-he-no-Wenah insisted he stay home and told him she had been warding off suitors while he was away.

In 1829, Prescott was hired to recruit and relocate Sioux families to an agricultural village established not far from Fort Snelling at a place named Lake Calhoun. This lake is now in the heart of Minneapolis and, in 2018, was renamed Bde Maka Ska. The goal of the institution employing Prescott was to convince the Sioux to cease their nomadic ways and embrace farming.

Prescott served as superintendent of the operation and was paid ten dollars per month. During the first summer, only two individuals participated in the project, but by the second summer, in 1830, Prescott had attracted about 250 Native Americans to the settlement.

Prescott was hired away from the agrarian program in 1832 by the American Fur Company, which was seeking to expand the company's capacity. Prescott's experience and skills finally paid off. He signed a three-year contract for the princely sum of $400 per year. And so, in the late summer of 1832, the company assigned Prescott to lead a troop of about one hundred people, including Sioux trappers and hunters, from near Fort Snelling westward into Indian country to establish a trading center and pursue fur trapping on the Big Sioux River. "I was ordered," explained Prescott, "to prepare for the Crooked River, a tributary of the Missouri."

Prescott's group departed during late summer and included his wife and family. They traveled cross-country using horse-drawn carts, eventually ascending the Coteau des Prairies. Here, they began seeing a new type of stone outcropped on shaggy grasslands. "We discovered a change in the color of the stone," Prescott observed. "[They] were of a reddish cast and the appearance of granite." It was likely the first time any of the party had seen the stone that later became known as Sioux Quartzite.

Prescott and the party arrived at the Big Sioux River, at a place historians believe was near today's Flandreau, South Dakota, and he selected a position for their settlement. All able-bodied souls began building a shelter as winter was approaching. The design of that structure was simple: house the multitude inside a single narrow building, some eighty feet long, and install as many partitions as were needed to provide a small amount of privacy for everyone. Trees were felled. Fireplaces and chimneys were constructed. And plenty of firewood was cut. By early December, the group was nestled into their tiny apartments.

The makeshift community was protected from the weather, but food, particularly meat, was another matter. In winter, game was scarce. Prescott's men killed several wolves, but they preferred eating hawks shot by Prescott. To find food, the hungry Sioux left the enclave, heading south toward the Missouri River. Prescott took several men to locate the Indians and entice them to come back to his post. In exchange, Prescott would give them cornmeal.

As Prescott and two associates tracked the Indians, they came upon the falls of the Big Sioux River. Historians believe Prescott was the first white man to document the cascades and the future site of Sioux Falls.

In 1832, frontier fur trader Philander Prescott became the first Euro-American to document an eyewitness observation describing the cascades of the Big Sioux River. *Public domain.*

Prescott described the falls as twenty yards wide with a drop of ten feet. They fall, he wrote, "through so many broken rocks and crevices that you cannot see much water about the falls when the water is low." There was no flourish or flavor in his words. And there wasn't much water to be seen. In a typical December, water flowing in the river and down the falls is greatly diminished from the river's average annual flow. Covered by snow with low or no flows, the cascades likely looked unremarkable. Prescott's impression might have been different had he visited in April or May.

Prescott and his men continued south but were unable to overtake the Sioux, so he turned back, camping again at the falls. This time, he wrote nothing in his journal about the place.

The winter of 1832–33 came and went, and few beaver pelts were procured. Spring's arrival exposed a shortcoming of the trading post site. Prescott had built the large structure too close to the river, and high flows forced everyone to tents on higher ground. Unfortunately, the onset of floods did not coincide with the departure of winter. Frigid winds and snowstorms battered the tent dwellers. It's no wonder that when cold weather transitioned to warmer spring, the group packed and left, vacating their spartan and short-lived post.

Prescott's fortunes reflected the unpredictable life of a fur-trading frontiersmen. During one busy winter, he cleared $1,000. In other years, he'd supplement his trading income by hauling wood, peddling corn, working as an interpreter for the military, guiding surveyors or leading missionaries to potential congregants.

One series of significant area expeditions with which Prescott did not assist was an ambitious cartographic undertaking led by Joseph Nicollet that intended to map the upper region of the Mississippi River's watershed. Nicollet, a distinguished French scientist and cartographer, had already explored and surveyed the southern portion of the Mississippi River basin when the United States government commissioned him to complete mapping the entire river basin. To finish the project, Nicollet and a small team carried out three expeditions during the years 1838 and 1839.

During his reconnaissance expeditions, Nicollet's party camped beside the Big Sioux River west of today's Brookings, South Dakota. "We pitch our tents," wrote Nicollet, "on the banks of clear, swift-flowing water meandering across an immense prairie whose vegetation is better supplied and more varied and where the land seems disposed to provide all the agricultural needs for civilized society." He described the prairie as growing roses and berries that "perfume the air and refresh the party."

Nicollet employed that era's cutting-edge scientific equipment and know-how to determine altitudes, longitudes and latitudes, and he included Native American place names on his maps, demonstrating his respect for the linguistics and ethnography of Native people. He made more than ninety thousand field observations to create his Mississippi River basin maps.

Nicollet's masterpiece, a grand map of the river's northern watershed, including modern-day eastern South Dakota, was titled "Hydrographical Basin of the Upper Mississippi River." Published in 1843, this map was especially notable for depicting the Coteau des Prairies.

In his report to the U.S. Senate, Nicollet described the coteau as an elevated plain, an expansive plateau that presented the highest elevation in that vast region between the Gulf of Mexico and Hudson's Bay. The view from the coteau's eastern margin that soared above glaciated lowlands in what is now Minnesota was, according to Nicollet, "magnificent beyond description."

A pair of significant channel bends in the Big Sioux River was identified by Nicollet, and the southernmost of the two, he said, was terminated by a cascades. These were the falls of the Big Sioux. A central purpose of Nicollet's explorations was to determine commercial opportunities within the upper Mississippi region. Navigation moving upstream from the Big Sioux's mouth on the Missouri River, explained Nicollet, could advance no farther north than the falls.

Nicollet's notes reveal that he did not personally view the falls, so his descriptions were based on input from others. Close examination of his principal map reveals that Nicollet misrepresented the course of the river as it flowed through what is now the Sioux Falls area. He more accurately identified a number of water bodies situated near the Big Sioux River north of the falls, including lakes that retain the names he used on his map, such as Kampeska, Punished Woman, Poinsett and Albert. These were places he personally visited. Nicollet also revealed that the source of the Big Sioux was within the northern reaches of the prairie coteau.

Nicollet titled the Big Sioux River "Sioux R" and also applied the Sioux name, Tchankasndata, meaning "thickly wooded river." In his journals and secondary sketches, Nicollet added the Sioux moniker Watpa ipha Jkshan, translated as "crooked river" or "river that bends." He explained that Native

American place names as applied to the same river might vary depending on a river's characteristics. "The Indians change the name of a river often along its course," said Nicollet. "Their geography gives information of immediate use to them. The names multiply because of useful objects, or memorable events, or formations of the terrain which are found along the river."

Nicollet applied additional names to the Big Sioux River, including La Rivière Croche (Crooked River) and Riviere des Sioux (River of the Sioux). His principal map did not acknowledge the location of the cascades of the Big Sioux.

Accompanying and assisting Nicollet on his expeditions was a soldier named John Fremont, a young man who impressed Nicollet with his knowledge of astronomy, geology and biology. Following his successful work with Nicollet, Fremont used his skills to create a map that depicted the entire length of the Oregon Trail. His colorful reports about the West were widely circulated and motivated countless Americans to venture to the frontier. He earned fame and a flattering nickname, the Pathfinder, and later parlayed that notoriety and experience into an impressive political career, including service as California's first U.S. senator and the Republican Party's nominee for the nation's presidency.

As Nicollet and Fremont charted the Northern Plains region, Philander Prescott was watching the world around him change. His time as a fur trader was ending. Rivers in the region were being ransacked of mammals, and Indian tribes that had developed a reliance on the pelt trade with whites faced a worrisome future. In 1843, Prescott accepted a job as an Indian interpreter and liaison for the federal government and the military operating out of Fort Snelling. He also oversaw farming programs for Sioux bands on seven different reserves.

In the summer of 1844, Captain James Allen and a command of fifty-six men from Fort Des Moines traveled north by horseback to find the source of the Des Moines River. A detailed summary and map of Allen's expedition was presented to the United States House of Representatives in 1846.

After Captain Allen successfully identified the river's source, he and his men traveled west and were excited to discover bison grazing lush prairie. A short time later, they reached the Big Sioux River, and they followed its course until September 13, 1844, when they arrived at the cascades. Allen penned the following in his journal: "[We came to a] great and picturesque fall of the river.…These falls present a remarkable feature of the river and country; the river, until now, running nearly due south, makes above the falls a bend to the west, and round to the northeast course for six miles, where it resumes its former direction." It is clear that Allen, unlike Philander Prescott, was impressed by the falls. "The fall," wrote Allen, "as near as I could measure it, is 100 feet in 400 yards, and is made up of several perpendicular falls—one 20, one 18, and one ten feet."

Allen was thrilled at the Sioux Quartzite massed around the falls. "The rock of these falls," he described, "is massive quartz.…The rock in the course and on the borders of the stream is split, broken, and piled up in the most irregular and fantastic shapes, and presents deep and frightful chasms, extending from the stream in all directions."

The party's mapmaker, Joseph Haydn Potter, had just graduated from the U.S. Military Academy, ranking number two among graduates behind his classmate Ulysses Grant. Potter's depiction of the Big Sioux channel showed the river abruptly turning north as it approached the falls. No earlier map so accurately located the falls as Potter's rendering.

An early treaty between the United States government and the Sioux people had been approved in 1825. Lands in the southern portion of modern-day Minnesota and throughout what are now North and South Dakota were designated Sioux homeland. Those protections were quickly disrespected, and between 1830 and 1837, the Yankton Sioux people relinquished 2.2 million acres of land. Additional treaties between the United States and the Sisseton and Wahpeton Sioux and the Mdewakanton and Wahpekute Sioux were completed in 1851, creating two small reservations on either side of the Minnesota River.

The Sioux people living east of the Big Sioux River had grown desperate. Hunting grounds had been invaded by white settlers, and food and trading resources were depleted. In exchange for the reservations, Sioux leaders surrendered millions of acres for a fraction of their true value. Some of the money was available immediately; some of it would be dispensed by government officials from a fund established by the treaty.

Eight years later, another treaty with Native people impacting Minnesota Territory was enacted. More than eleven million acres, including lands in what is now eastern South Dakota, were ceded to the United States. In exchange, the Yankton reservation was created, and the Yanktons were recognized as the caretakers of the Pipestone quarry.

Treaty negotiations were complex and acrimonious, and Sioux leaders were deceived and misled. There were assurances that necessary supplies, including food, would be provided. Those promises proved to be worthless. Hunger and deprivation characterized life for the Sioux on the reservations.

Philander Prescott participated in at least one treaty negotiation as an interpreter, and he described the process as uncomfortable. In a letter to the Bureau of Indian Affairs, Prescott complained that the government was not satisfying its obligations and payments to the tribes. "The treaties say these funds shall be annually expended, whereas large amounts have been kept back and are now in arrears," wrote Prescott. Government officials asserted that Prescott was in league with the Sioux.

In about 1852, Prescott and his family secured one of the first land claims in what would become the city of Minneapolis, and he built a large home there. Ever entrepreneurial, Prescott opened a sawmill to serve the steadily growing city. At that point, he and his wife had been married nearly thirty years and had parented a large family. The veteran frontiersman joined the rising class of accomplished residents to celebrate the civic culture that was budding in the new community. He also continued to work with the government on matters related to Indians.

In his final report about the educational farming operation he'd overseen, dated 1856, Prescott described the unhappiness of the Sioux living on supervised reservations. "We have had more trouble with them the past three or four years than I have known in thirty-four years I have lived with them," he reported.

Chapter 3
SUNSET LAND

Thomas Jefferson and Alexander Hamilton vigorously debated the superior approach to distributing the nation's publicly owned western lands. Jefferson advocated for egalitarian policies favoring the common man by offering land at affordable prices. Hamilton felt the nation would be better served by encouraging the wealthy and well-connected to acquire property and then resell it to settlers. Jefferson's philosophy prevailed, though opportunities for the investor class were included in policies intended to help developers exploit ripe new territories.

An especially important strategy to settle the West was the 1862 Homestead Act, a promotion that granted settlers 160-acre parcels for twenty-six dollars. Homesteaders using this law were to be at least twenty-one years old or the head of a household. In exchange for inexpensive land, a settler had to build a home and make other improvements, including planting crops and satisfying a five-year residency requirement. The act was later amended to preclude Confederate soldiers from the program and reduce the number of years Union soldiers were required to live on a homestead by the amount of time they served in the Civil War. A progressive feature of the homestead program allowed former slaves, women and immigrants to possess a homestead.

Homesteaders took a leap of faith to settle isolated lands. Those with bigger visions, like town building, jumped even farther. Business speculators benefited from an 1844 law that offered town builders up to 320 contiguous acres. This provision encouraged a small but spirited town-building industry.

Inventing a town from scratch to profit from its growth was a risky business that involved many factors. Which settlements would thrive, and which ones would wither? What would be the chosen route of a stagecoach trail? Could railway service be secured? What about mail service? It was practically impossible to plan contingencies to cope with the countless threats to a speculation settlement. A certain level of high-minded bravado was necessary. Across today's Northern Plains are a handful of towns that became cities and hundreds of hamlets that were stunted or vacated.

In 1856, a book intended to ignite public interest in settling the western frontier was published to much acclaim and attention. Titled *The States and Territories of the Great West*, the book served as an inspiration for settling Sioux Falls. In the introduction to the book was the following: "No parallel can be found in the world's history to the progress and the prospects of the Great West."

Author Jacob Ferris had traveled and studied every part of the region stretching from Ohio to the Rocky Mountains, and he wrote eloquently about opportunities for a man to make a new life for himself in this raw, enigmatic expanse of America. Ferris noted settlement and townsite opportunities, as well as general geographic information. He was impressed with the scenic setting and industrial potential where Sioux Falls would one day bloom. "The Big Sioux River," he wrote, "breaks through a remarkable quartz formation, and seems to have ruptured the massive wall of rock.…Within a distance of four hundred yards, the river leaps and plunges down three successive falls…with rapids intervening, supplying an incalculable amount of waterpower." Quartzite, a high-quality building material, was plentiful at the falls and easy to access, explained Ferris. "The stone," he declared, "is not affected by acids and is said to be indestructible by fire."

A Dubuque, Iowa physician and businessman named George Staples devoured Ferris's book and launched a town-building enterprise called the Western Town Company that immediately began considering settlement sites in Iowa, Nebraska and Minnesota. The setting alongside the falls of the Big Sioux River in Minnesota Territory seemed to be a sure winner, and Staples and his associates pounced.

Staples sent two agents, Ezra Millard and David Mills, to find the falls and stake a claim there. From Sioux City, the men proceeded north, traveling in a horse-drawn wagon loaded with supplies. Because much of the river was bordered by steep slopes, ravines and impassable timber, the two pioneers followed the waterway high above the channel's eastern shore on smooth blufftops, where trees gave way to grass. After ten days of careful travel,

Author Jacob Ferris's glowing descriptions of the cascades enticed townsite speculators. *Courtesy of Robert Kolbe Dakota Collection.*

they spied their prize, the falls. "They realized," wrote historian George Kingsbury, "they had found one of Nature's grandest marvels that would become famous among the scenic splendors of the world."

To initiate Western's claim, the two men built a small log cabin on a timbered island at the head of the falls. It appears Millard left for the winter and Mills remained or returned to Sioux City. Millard likely never did return to the falls, though he remained a regional pioneer, serving as mayor of Omaha, Nebraska, from 1869 to 1871. Mills eventually farmed in nearby Iowa and later moved across the Big Sioux River to Elk Point, Dakota Territory, where he was elected to the territorial legislature.

In the spring of 1857, another Western employee, J.T. Jarett, arrived to survey and plat the land and begin additional improvements to the claim. He supervised a team of ten Western employees who hauled provisions and supplies to the site and began building a new town. The group was well-armed and worried about local Indians. They immediately assembled a sawmill, and nearby timber was felled and trimmed. Several wooden houses and a store were constructed and a stone structure, too. Among the town's first pioneers were friends Wilmot Brookings and Dr. Josiah Philips, both Maine natives.

Brookings was a resourceful man compelled by grand ambitions. He'd earned an undergraduate degree from Maine's prestigious Bowdoin College in 1855, where George Staples earned his medical degree. During his time at Bowdoin, Brookings met Phillips, a young doctor. Phillips's father, also a physician, was George Staple's partner in the Dubuque, Iowa medical practice. Although Brookings's experience was as a high school teacher, he gravitated to the study of law and was admitted to the Maine bar in 1857. He then traveled west, as many adventurous New Englanders before him had done. His objective: open a law practice in Iowa. After only seventeen days in Dubuque, Brookings took a chance on wilder territory. He and three others representing the Western Town Company, including Josiah Phillips, departed company headquarters bound for the Big Sioux River valley. They added six men to their outfit in Sioux City.

During the long trip from Dubuque, Brookings's duty was to oversee the oxen pulling supply carts, a job commonly described as a bullwhacker. He would later be promoted to managing agent for Western's venture at the falls.

Western's holdings included land north and west from the falls including the island, and the company titled its new settlement Sioux Falls.

A place so appealing was bound to attract attention, and soon, a second town-building enterprise appeared. From St. Paul, Minnesota Territory, came the Dakota Land Company, a well-financed, well-connected enterprise determined to identify and develop townsites in what is now southwestern Minnesota and southeastern South Dakota, including along the Big Sioux River.

It appears that during an 1856 exploration of potential townsites, agents representing the Dakota Land Company visited the falls of the Big Sioux River but didn't encounter their rivals from Iowa. It is possible their visit preceded Western's arrival. They claimed no land at the falls but returned to Saint Paul with a recommendation to do so.

The first Sioux Falls settlers admired a vast, untouched landscape surrounding the wild cascades. *Courtesy of Robert Kolbe Dakota Collection.*

The Dakota Land Company's scheme was more complex than the strategy pursued by the Western Town Company. The St. Paul syndicate anticipated that as Minnesota gained statehood, parts of the old Minnesota Territory not included within the boundaries of the new state would become a new territory. The Minnesotans had strong political influence, and they aimed to gain control over the new territory and guide decisions regarding locations of counties, the capital, a public university and a penitentiary.

The ambitious intentions of Dakota Land emanated from the top of the business. Samuel Medary, who had come to St. Paul from Ohio in April 1857 to serve as the third governor of Minnesota Territory, was involved in the company's formation and supervision. Medary had already led a controversial life as a newspaperman and Ohio state legislator and as an unrelenting firebrand within the state and national Democrat Party. He was offered an appointment as territorial governor because of his enthusiastic support for candidate James Buchanan during the previous year's intense presidential election, when Buchanan defeated Republican John Fremont and third-party candidate and former president Millard Fillmore. Fremont, a Georgia native, was a fervent slavery opponent, in contrast with Buchanan, who aligned himself with the so-called Peace Democrats, a contingent of Americans who opposed resorting to war to settle the slave dispute. Buchanan's platform advocated for each state making its own decision about slavery, and he prevailed in every slave state except one to secure his victory.

Fremont had risen to the national stage because of his role encouraging settlement of the West. He became a founder of California's Republican Party, served his state in the U.S. Senate and was the Republican Party's first candidate for president. Eighteen years before his unsuccessful bid to occupy the nation's highest office, Fremont helped Joseph Nicollet explore and map the northern reaches of the Mississippi River watershed, and during the summer of 1838, the two men camped beside the Big Sioux River some forty miles north of the cascades.

Managing a designated territory was a plum position, with access to information and influence over territorial decision-making. Medary and other high-placed politicians, merchants and businessmen pondered town-building opportunities in Minnesota and the entire western region. An insider's influence, they knew, would be invaluable. Medary and his allies in Minnesota successfully manipulated the establishment of a critical border of the new state that favored the Dakota Land Company. An original plan for statehood fixed Minnesota's southwestern border at the Big Sioux River. That border was shifted east, so the Big Sioux would lie in a new territory lacking formal government and more easily manipulated by town developers.

Unresolved were Indian issues. Easing some worry in the Big Sioux valley was the anticipated formalization of a treaty with the Sioux that would open the area to white settlement. But when the Dakota Land Company began planning its western excursions, that land was still Indian Country, and white settlers would be trespassers.

It was no coincidence that when the Minnesota territorial legislature convened in 1857, a group of ten counties was officially created in what would become the southwestern corner of a new state named Minnesota and in lands farther west in an area that would become Dakota Territory. These new counties would need seats of government, and six of the townsites chosen by the Dakota Land Company were tapped by lawmakers as future county seats. "It was the fortune of our Company," stated unsurprised Dakota Land officials, "to secure the Seats of Government to those Counties." The company's grand plan was off to a promising start.

The town builders quickly undertook the next phase of their blueprint. Sites already identified had to be formally claimed, and occupants needed to commit to one year of service to the company. Thirty-five hardy souls volunteered to serve the town-building mission. A year's worth of supplies, including agricultural and mechanical implements, were packed. So were weapons, ammunition and construction materials.

On May 21, 1857, the small group of town-building pioneers, including several members of the Dakota Land Company's board of directors, departed St. Paul to considerable fanfare. Streets were lined with applauding onlookers, and banners saluting the Dakota Land Company were displayed. A newspaper celebrated the company's noble intentions: "A praiseworthy object on the part of men…in an effort to subdue and open up another garden for the home and happiness of the great transient family, still without lands and without homes." Opening "uninhabited" lands to settlement was viewed as the work of heroic figures.

Like a legion of valiant conquistadors heading off to war, the town builders paraded out of St. Paul to a waiting steamboat. That vessel carried the pioneers and supplies up the St. Peter's River (the Minnesota River) to the Upper Sioux Indian Agency. Word of the glorious undertaking had spread. As the boat passed communities along the river, hundreds stood on the shoreline to cheer. At the agency, the group set ashore to begin the next phase of their assignment. At least three separate teams began their overland marches to predetermined prospective townsites.

One of those sites, named Saratoga, had already been designated the county seat of Cottonwood County, Minnesota, and was located sixty miles west of New Ulm. At this site, the Dakota Land Company declared, "There are several copious springs…of highly valuable medicinal properties."

A second townsite, titled Mountain Pass, was situated at the head of Lake Benton. It was expected that a critical mail station on a new government road linking Minnesota with the West Coast would be located at Mountain Pass.

Medary was the third town, and it would be situated along the Big Sioux River twenty-five miles west of Mountain Pass. At this townsite, named to honor Samuel Medary, a proposed government road would ford the river. Medary would serve as the county seat of Midway County, Dakota Territory.

The company also acquired land at Flandreau, the proposed county seat of Rock County, Dakota Territory, located about fifteen miles south of Medary.

Downstream from Flandreau were the cascades of the Big Sioux River. Company representatives arriving there were surprised and disappointed to discover that a rival town company had already claimed land near the falls. Nevertheless, Dakota Land staked a claim for 320 acres south of and adjacent to the property already claimed by the Western Town Company and named their town Sioux Falls City. This site, containing much of what would eventually become the downtown district of Sioux Falls, was considered the crown jewel of all Dakota Land Company prospects, and it was anticipated that Sioux Falls City would be named the county seat of Big Sioux County, Dakota Territory. Company officials also intended the community to, one day, become the territorial capital.

Reports indicate that the two town-building rivals comfortably coexisted. There was, it seems, plenty of opportunity to share.

During the autumn of 1857, George Staples traveled across Iowa to visit his company's newly sprouted townsite at the falls of the Big Sioux River. He was delighted to learn that Jacob Ferris hadn't exaggerated the site's potential. "There is no place yet known within the territory of Dakotah," said Staples, "that has so great natural advantages as Sioux Falls." Staples described the main section of the falls: "The waters begin to tumble over the successive strata of rock as they cropout, for the entire bed and banks of the stream for two thousand feet along the Falls above and below, are nothing but the everlasting rock." He likened the series of waterfalls within the long stretch of rocky rapids to an immense stone stairway, with the final drop in the series of steps as the tallest, at twenty-one feet, and he cheered the river's strong current as a potential source of energy.

Minnesota became a state in 1858, but Indian troubles accompanied that momentous happening. Skirmishes had been occurring for some time, with the most serious in 1857, when Sioux riders attacked white settlers near Spirit Lake, Iowa, and killed almost forty people. By 1858, tensions had reached a fever pitch. Portions of established Sioux reservations had been wrongly opened to settlement, allowing farmers and woodcutters onto lands that had been set aside for the Sioux. Angry Sioux warriors, squeezed from their small reserves, overpowered the newly planted communities of Medary and Flandreau, demanded that all residents leave and then sacked and razed each town.

When a war party arrived in Sioux Falls, they discovered the city's three dozen inhabitants were well-protected, armed and prepared to fight. A ten-foot-high earthen barricade and a deep ditch on the outer portion of the protective barrier had been quickly erected and excavated. Within the dirt walls, an area measuring about eighty by eighty feet, was a quartzite stone building constructed earlier by the Dakota Land Company. Other buildings were left unguarded beyond the perimeter of the fort. The Sioux warriors assessed the crude yet formidable fortification and contemplated combat. After a three-day standoff, the Indians moved on. For six weeks, Sioux Falls settlers nervously clustered inside their small fort, aptly called Fort Sod.

By the time the Dakota Land Company released its first annual report to shareholders, in 1859, Samuel Medary had departed Minnesota to serve as territorial governor for Kansas. Medary's life took a turn for the worse when he returned from Kansas to Ohio. Because of his outspoken opposition to the Civil War, Union soldiers destroyed his newspaper offices. The federal government then charged him with conspiracy against the United States. While awaiting trial, Medary died at the age of sixty-three. Before his death, he symbolized resistance to a campaign waged by the federal government against newspapers opposing the war. Medary's connections to Dakota quickly faded, and the only notable homage to the man is the name of a busy street in Brookings, South Dakota.

The Dakota Land Company's 1859 report was surprisingly upbeat, only briefly noting the destruction of two of its settlements. Despite those setbacks and a siege on its most treasured townsite, promotions spread by the company continued a ceaseless drumbeat of favorable publicity.

The unsettled territory's promise was extraordinary, the company declared in an advertorial published in the *St. Paul Times* newspaper, explaining that those who visited the area "had modestly described [it] as one of the most beautiful and fertile regions in the sun-set land."

The bright future of Sioux Falls City, wrote the company, was confirmed when a group of dignitaries from St. Paul visited the village and falls. Dakota Land officials reported,

> *The cataract's loud laughter greeted our ears some time before we reached the brow of the bluff bordering semi-circularly the broad, beautiful vale down which we could look upon the Capital City and the great waterfall in its midst. On arriving at a point where a full view could be had, there was a general halt—a moment of silent awe—then a simultaneous shout, and cheer after cheer went up from the whole party. It was a glorious*

sight. A city barely two years old, flourishing with… its stream and water mills, its mercantile houses, mechanics shops and fine residences, county buildings, city offices, legislative halls…Along in front flows a copious and beautiful river. Opposite the center of the city occur the falls whose scenery is picturesque beyond description.

The overstated public relations piece compared Sioux Falls cascades to the best-known waterfalls in Minneapolis. "Minne-ha-ha is sublimely enchanting," company officials declared, describing the small, exquisite falls on Minnehaha Creek in what is now southern Minneapolis. "St. Anthony Falls [on the Mississippi River in what became downtown Minneapolis] is mighty and not without its picture, but you will have to combine the two to make a likeness of Sioux Falls, and even then the picture will be wanting. The skillful hand of nature seems to have excelled itself here."

Chapter 4

ANOTHER REBOUND

After the Sioux hostilities of 1858 were suppressed, people began trickling back to Sioux Falls. A scattered collective of wooden shanties and a lone stone structure constituted the material existence of the humble hamlet. The *Dakota Democrat*, based in Sioux Falls and associated with the Dakota Land Company, became the first newspaper printed in Dakota Territory. Among the items featured in the inaugural issue, published on July 2, 1859, was a prosy piece written by Henry Masters, a lawyer who'd helped organize the Western Land Company. Masters moved to Sioux Falls from Dubuque, Iowa, in the summer of 1858, and in September that year, he was elected to serve as provisional governor of Dakota Territory. The election was rigged by Sioux Falls partisans organized to inflate support for Masters.

In his commentary about the falls of the Big Sioux River, Masters urged restraint and counseled Sioux Falls residents to protect rather than exploit the cascades. "Here at the falls," wrote Masters, "a city shall rise and shalt learn that seeking gain is not which the sons of men most prize."

Not long after issuing his advice, Masters was dead from cerebral hemorrhage or stroke, becoming the first white person to die in Sioux Falls. Samuel Albright, owner and editor of the *Dakota Democrat*, filled in as governor until Wilmot Brookings was appointed to serve in the post. At that time, Dakota Territory was nothing more than an unorganized, orphaned tract that had been discarded when Minnesota gained statehood and Minnesota Territory ceased to exist. Ruling over this vast territory were freewheeling residents angling for power and opportunities to benefit themselves and their

Sioux Falls leader Henry Masters warned his community in 1859 not to sacrifice the beauty of the cascades for the needs of industry. *Courtesy of Robert Kolbe Dakota Collection.*

communities. The messy, clumsy political situation was derisively labeled a "squatters' government."

Samuel Albright used his newspaper and political platform to spread rosy messaging intended to lure new settlers to Sioux Falls. As 1859 drew to a close, slightly more than forty brave souls called the community their home.

The year 1860 delivered bad news for the Dakota Land Company. Abraham Lincoln's election to the presidency delayed official establishment of Dakota Territory until the spring of 1861 and shifted favorable territorial appointments, policies and laws away from the Democrats who steered the Dakota Land Company. The St. Paul company faded away.

In August 1862, Sioux Indian frustrations resulting from inefficient and dishonest American government again exploded. This deadly affair would last longer and spread through more of the region than the conflict four years earlier.

The backstory is straightforward. Government officials at the Lower Sioux agency refused to distribute warehoused food supplies to famished Indians. Those authorities determined that until the Indians were able to purchase provisions, their food would be withheld. But treaty payments due the Sioux were delayed. One desperate Sioux man stole food for his family from a nearby farm, and a posse of white settlers apprehended, murdered and decapitated him. A government agent named Andrew Myrick declared that until the Sioux paid, they could "eat grass or their own dung."

A Sioux war council decided that whites must be driven away. Warriors attacked Lower Agency on August 18, destroyed the village and killed Andrew Myrick. Apparently, he was found beheaded, with grass stuffed into his mouth.

Indian friends warned Philander Prescott to hide in his Upper Agency quarters until the danger of marauding warriors had passed. But Prescott opted to join a party fleeing to Fort Ridgely. The group was ambushed, and Prescott and others were killed. There was irony, of course, in Prescott's murder: a man who stuck up for the Sioux at the risk of his own reputation had been slain by the Sioux. Prescott, the first white man of record to view and describe the cascades of the Big Sioux River, was one of the first white victims in the 1862 uprising. In the days following Prescott's death, deadly chaos from the growing war quickly spread from the Minnesota River valley across the Northern Plains.

A group of Sioux braves attacked an Iowa settlement and killed fifteen settlers. The warriors then split up, with one group heading to Minnesota and the other to Sioux Falls.

On August 25, a war party arrived on a bluff overlooking Sioux Falls. There they encountered and killed two local residents, Joseph Amidon and his son. The elder Amidon was a farmer and local judge. From that elevated vantage, the Sioux looked down at a place—the cascades of the crooked, timbered river—that had once been a touchstone for their people. The view that greeted Indian visitors for many years had been altered and stolen. Revenge seemed appropriate, justified.

The army had stationed three dozen soldiers at Sioux Falls, and a detachment was sent to find the Amidon killers. Shortly after the soldiers departed, Sioux warriors attacked the village. Those soldiers remaining in town, assisted by civilians, fought back and were soon reinforced by the returning search party, forcing the Sioux to withdraw.

Three days later, the first official governor of Dakota Territory, William Jayne, issued an evacuation order to those in Sioux Falls, and nearly all the town's panic-stricken residents rushed to Yankton, the territorial capital. The Sioux returned when the village was empty to pillage the place and torch most buildings.

Lasting six weeks, the so-called Sioux Uprising of 1862 cost the lives of as many as 800 settlers and soldiers and 150 Sioux. Settlements and farms were plundered and destroyed. The military apprehended 302 Sioux men who allegedly killed whites during the brief war and sentenced each to death. President Lincoln commuted the death sentences of 264 prisoners, and 38 Sioux were subsequently executed by hanging in Mankato, Minnesota, the largest mass execution in the history of the United States. Three years later, 2 Sioux warriors thought to have murdered Philander Prescott and his companions were hunted down in Canada, brought to Fort Snelling and hanged from a gallows not far from where Prescott and his family once lived.

In October 1864, a small group of soldiers and surveyors on a mission from Fort Randall, Dakota Territory, arrived at deserted Sioux Falls while mapping eastern Dakota and the Iowa border. According to Moses Armstrong, a member of the surveyors' crew, the countryside surrounding Sioux Falls teemed with buffalo, wolves, deer and coyotes. On a previous visit to the community, in 1859, Armstrong claimed he could hear the roar of the falls from a distance of three miles. The scene that greeted Armstrong in 1864 was bleak. Sioux Falls was a ghost town, looking dusty, weedy, forlorn and eerie. Most structures had been destroyed, and pieces of household items were strewn about.

Moses Armstrong and others bunked in a sturdy quartzite structure that had not been destroyed. It was the building that formerly housed the town's newspaper office, and at least one account indicates there may have been parts of a printing press still standing in the main room.

Sioux Falls developed an especially keen rivalry with Yankton. During meetings to determine the territorial capital, adversarial legislators fought and flashed revolvers and knives, forcing local soldiers to provide security and prevent bloodshed. Armstrong and his Yankton allies prevailed, lifting the fortunes of their river town. Only months after losing the capital competition, Sioux Falls was destroyed by marauding Sioux warriors.

Citizens wanting to return to Sioux Falls sought a military presence in the community, but their requests were unsatisfied until the Civil War concluded and troops were available. The likelihood of Indians attacking the settlement was small, but the 1862 conflict cast a broad, cautionary shadow, and protections were viewed as necessary.

The first several buildings supporting Fort Dakota were finished in the summer of 1865 and were situated not far from the falls and west of the Big Sioux River, on the site of present-day Phillips Avenue and Seventh and Eighth Streets. Among the eighteen structures that eventually comprised the installation, there was a circular stone tower, some fifteen feet tall and thirty feet in diameter. Mounted atop that tower were two twelve-pound howitzer cannons. The fort's secured property measured one thousand feet by two hundred feet and hosted a small force ranging from twenty to eighty soldiers. The army also established a seven-by-ten-mile military reservation surrounding the fort that prevented civilian development near the falls.

As Fort Dakota grew, so did grumblings by the same town leaders who had initially lobbied for a military presence in their community. Not only were residents confident that threats from Indians had waned, but they also wanted access to the choice real estate occupied by the army. It was during this time that the Western Town Company folded, joining fellow Sioux Falls originator Dakota Land Company as vanished stepping stones in the community's history.

Sioux Falls civilian promoters got their wish when the fort closed in 1869, after providing a protective deterrent for only four years. Opening military property to nonmilitary development elevated the prospects for an aspiring community.

There was considerable political maneuvering to determine how the newly available land and military buildings would be apportioned to private citizens. A group of Boston speculators lobbied for a bill that passed the U.S. House of Representatives. If endorsed by the Senate and signed by the president, the Bostonians would be permitted to purchase the old army property for only $1.25 per acre. But Sioux Falls activists fought back. A petition opposing the bill was quickly circulated, and Sioux Falls resident Richard Pettigrew personally delivered the document to Wisconsin Senator Matthew Carpenter. Earlier, the Bostonians had bribed Pettigrew to abandon the petition drive, offering him ten sections of land. Pettigrew rejected the inducement and cheered as the Senate denied the land sale.

When the federal government officially vacated its Fort Dakota holdings, there was an auction to sell the former fort's buildings and land, and Dr. Josiah Phillips acquired seventeen buildings and at least some of the property that had hosted the military post. Phillips later broke up his valuable tract by selling lots to various citizens. This arrangement spread lucrative real estate to a number of local entrepreneurs, although Phillips shrewdly retained the sizeable parcel that would eventually become Sioux Falls' core commercial district.

Wilmot Brookings likely acquired some of the Fort Dakota property, including land near the falls and the island located a short distance upstream from the falls, one of the area's most unique and lucrative pieces of real estate.

Brookings's doughty character and outsized role in the development of Sioux Falls endeared him to the community. An earlier episode in his life boosted his mystique. There are several versions of this story, but the most likely account, as reported in a Sioux City newspaper, suggests that in early February 1958, Brookings and a business partner named Kinsey left Sioux Falls on horseback, aiming for Sioux City's federal land office, where they planned to begin the process of acquiring land and establishing a townsite on the Missouri River. The weather was wintry and forbidding, making it a questionable time to travel. When the two men encountered an icy, swollen Rock River, about fifty miles south of Sioux Falls, they turned back. By then, Kinsey had fallen ill and was struggling to travel. About fifteen miles from Sioux Falls, battling a blizzard and the coldest temperatures of that winter, the two men attempted a shortcut that required crossing the Big Sioux River. Tragically, Brookings slipped into the frigid water. Drenched and cold, he weighed his options. After making his suffering companion as warm and comfortable as possible, Brookings set off on foot for Sioux Falls, a trip that must have been grueling and torturous. By the time Brookings reached Sioux Falls, his feet were frozen inside his boots. Josiah Phillips was summoned to tend to his friend, and the young doctor made a dire diagnosis: both of Brookings's legs must be amputated above the knee. The procedure was performed in a drafty, candlelit cabin using a large butcher knife and a small saw. Though unspeakably painful, that crude surgery saved Brookings's life.

After six months' bed rest, Brookings traveled by cart and railroad to Philadelphia, where he was fitted with wooden prosthetics. By early 1859, he was back in Sioux Falls, using a cane and pursuing business deals and public service with his customary zeal. Later that same year, he was elected to the territorial legislature and was named acting territorial governor when Henry Masters died unexpectedly. Brookings also served on the territorial supreme court and ran a local newspaper, the *Sioux Falls Leader*. The city of Brookings, South Dakota, and Brookings County were named after him. So was the small island he owned near the falls of the Big Sioux River.

Between 1871 and 1873, Sioux Falls' population doubled to six hundred residents. The fledgling community was becoming a center of services and commerce. Quarries mining Sioux Quartzite were started. Rural residents, farming settlers, began to fill the countryside around Sioux Falls. Included among the fifty-four wood or stone buildings were thirty homes, three hotels, numerous businesses, a post office, one church, several newspapers, a bookstore and a brickyard.

Sometime in early 1872, a vagabond writer named Amos Gottschall arrived in Sioux Falls. His visit was purposeful, yet he had no firm plan. He would do in Sioux Falls what he had already done and would later do in many other locations: find short-term employment to pay for lodging and sustenance as he learned about a new place and the countryside around it. His was a twelve-year adventure across every region of the nation, no matter how remote, and through all major cities.

Gottschall's account of his odyssey was presented in a book titled *Travels from Ocean to Ocean and from the Lakes to the Gulf*, published in 1884. His explorations, he wrote, ranged from "Indian camps…to the hut of the hunter and pioneer, and in the parlor of civilization."

Gottschall's wanderlust began at age ten, with a long, unchaperoned hike through the countryside near his home in Marietta, Pennsylvania. His worried parents searched far and wide for two weeks until little Amos returned on his own accord, unharmed and emboldened.

On another early journey, Gottschall found himself on a wharf in Baltimore begging the captain of an English ship to take him to sea. The mariner advised the young lad to go home to his mother. "Early in life," wrote Gottschall many years later, "the desire to leave my home and see the world and its people outside of my quiet, narrow sphere was the leading theme of my childish thoughts and plans.…My thirst for travel and adventure was part of my very existence."

At age sixteen, Gottschall, already a seasoned traveler, launched his most ambitious journey, an adventure he called his "ramble." It appears he learned the craft of typesetting at a Chicago newspaper, and that skill would subsidize and propel his travels. Throughout his national exploration, Gottschall logged time at thirty-seven newspapers and printers in a dozen states, including two newspapers in Sioux Falls.

During a visit to Minneapolis, Gottschall found work at a printing company, squirreled away his paychecks and then traveled to Elk Point, Dakota Territory, located near the mouth of the Big Sioux River. His ultimate destination on this leg of his journey was Sioux Falls, sixty miles

This page: The natural cascades inspired poetry and flowery prose. *Top image courtesy of Siouxland Heritage Museums; bottom image courtesy of Robert Kolbe Dakota Collection.*

north. He reported that the walk from Elk Point to Sioux Falls, following the valley of the Big Sioux River, took him through "one of the most enchanting prairie districts the sun ever shone upon."

When Gottschall first stepped into Sioux Falls, the community's notoriety revolved around the wild cascades, and the young newcomer was impressed with the scenery. He wrote,

> *The Sioux River, a very quiet prairie stream, suddenly awakens from its wonted lethargy and calmness, and nearing a succession of massive red boulders, becomes one dashing, whirling, roaring body of spray and foam, and with a tremendous, rushing sound, leaps over the rocky chasms, thus forming the picturesque, romantic and fascinating Sioux Falls.*

Gottschall discovered that Sioux Falls possessed a rising culture and strong community spirit. The village was coarse but also gracious, despite being situated on the eastern edge of the western frontier. Sioux Falls might have been a border town, but it didn't behave like one. There were no gunfights at high noon or vigilante mobs.

A Sioux Falls tale popular among history buffs involves the fate of a printing press that served the community's first newspaper, where Gottschall found employment. This press may have been used in the same building where Moses Armstrong and others stayed for a short time eight years earlier. Gottschall later claimed that part of the same machine he operated had been flung onto rocks along the Big Sioux River near the falls by Sioux warriors during their 1862 invasion. Various versions of the story disagree about the amount of harm done to the press and who actually perpetrated the damage.

One entertaining assertion explains that when the Sioux plundered Sioux Falls, some of the letters and numbers that had been part of the original press were confiscated and later integrated as ornamental features onto pipes and other items hand-fashioned from pipestone. According to early Sioux Falls pioneer Richard Pettigrew, workers discovered a printing press platen along the river. This flat, heavy piece of iron measured nearly twenty-eight inches by twenty-two inches, was one inch thick and weighed about one hundred pounds. It had been used in the printing press as a plate to impress inked letters and numbers onto paper, and it is now stored at a museum in Sioux Falls.

Perhaps the platen had malfunctioned before the Sioux attacked and had been casually discarded where many residents dumped refuse and other items. Or perhaps someone other than Indians vandalized the press. If the Sioux in fact sabotaged the printing press, it was a fitting form of comeuppance—although it's doubtful the marauding Indians realized the role of this press in attracting settlers. Had they understood, they might have obliterated the apparatus, leaving nothing to be recovered and repaired.

Another curious aspect of the printing press's tale traces its improbable itinerary. Storytellers claimed the Gutenberg contraption was subjected to a journey that rivaled Gottschall's. Built and first used in Cincinnati, Ohio, the press was later shipped to Dubuque, Iowa, where it printed the first newspaper in that state. Its next destination was St. Paul, Minnesota, where the press was used to publish the *Pioneer Press*, Minnesota's inaugural newspaper. The Dakota Land Company then acquired and transported the machine from St. Paul to Sioux Falls, where it printed the first newspaper in Dakota Territory. Although there is an appealing charm to that narrative, historians at the Minnesota Historical Society dispute it, claiming the press shipped from Iowa to St. Paul never left St. Paul.

The primary reason Gottschall drifted to Sioux Falls was the town's proximity to Plains Indians. Gottschall's fascination with Native Americans stemmed from a tragic family story. In 1745, his great-grandmother, at age seven, was abducted by Indians after they killed her parents and sister. The little girl eventually escaped and fled to a white settlement. Years later, young Amos read stories about the 1862 Sioux uprising. He read about tepees and buffalo hunts and eagle feathers. "The stories I had heard about the Indians and their wild, nomadic ways," remembered Gottschall, "so impressed my young mind that I resolved to see and know these people."

It was a grim time in America when Gottschall went looking for Indians. Many white citizens urged extinction of the continent's Native people. It was widely believed by whites that there were no lessons to be learned from Indians and that their resistance to the ways of those who'd overpowered them was impertinent and bad-mannered. An optimistic, impartial soul, Gottschall sought the rest of the story. And so he walked alone into Indian villages, befriending Native American people, often staying with gracious, generous families inside their lodges. He politely asked questions and took notes about events and people. His desire to learn the Sioux language brought him respect and affection.

"I used to tell the Sioux," wrote Gottschall, "that I intended to make a 'wo-wah-pe' [book] for the white people to read, and in it I would describe the Indian mode of life and inform my people just how I fared at the hands of the Indians. This always pleased them and doubtless influenced them to treat me well." One host in a Sioux village, a fellow who had participated in the 1862 conflict, asked Gottschall to teach him to read and write English so he could recognize and thwart the lies heaped by whites on Indians. More than once, Gottschall smoked the pipe with a warrior who had slain *wasichu*.

Amos Gottschall quietly departed Sioux Falls, continuing his grand journey. The little town he left behind commenced its own new adventure. Modernizing was afoot. A fresh era was underway. Statehood lay ahead. Settlers were streaming into the region. And the public's posture toward the cascades was changing. Gottschall and other early visitors described the falls using spirited, poetic expressions. Growing, ambitious Sioux Falls would soon embrace a more commercial attitude toward its namesake. The restraint advised by early community leader Henry Masters would be disregarded.

Chapter 5

INDUSTRIALIZATION

In his first speech to Dakota Territory's House of Representatives, on March 17, 1862, Governor William Jayne boasted about the commercial potential of the cascades of the Big Sioux River. "The falls on the Big Sioux River," Jayne declared, "furnish a mode of power sufficient to drive any of the machinery used in New England mills."

At that moment, New Englanders were operating thousands of mills with varying capacities and capabilities. In addition to gristmills, which ground grain into flour, there were paper mills and sawmills. Metals were shaped, polished and sharpened. Ore was crushed, and yarn was spun.

Colonial Americans began erecting water mills soon after arriving in the New World. The practice of converting moving water to power traveled with those first settlers from Europe, where waterpower had been used since before the Middle Ages. A gristmill built in Jamestown, Virginia, in 1621 has been identified as the earliest American water mill. By 1850, there were over one hundred thousand functioning water mills in the United States. George Washington ran a gristmill at his Mount Vernon farm that featured an indoor waterwheel, and he later installed the first automated milling system. In the realm of millers, Washington was considered an innovator. Few towns in New England lacked a water mill, and many towns and cities had multiple businesses using their own water mills. A community with a water mill proudly referred to itself as a mill town, and the folks in Sioux Falls coveted such a moniker.

Those easterners who first settled Sioux Falls could hardly be blamed for their expectations regarding the falls. The potential of the falls to generate homegrown energy dwarfed prospects found at other communities in the region. The area surrounding Sioux Falls was fertile and agricultural; using waterpower to grind grain would provide a convenient market for farmers; and area residents could easily access flour, a nutritious food staple.

Historians identify the Webber-Hawthorne facility, opened in May 1873, as the first large gristmill to exploit the falls of the Big Sioux River. Apparently, a number of smaller waterpower businesses already existed on the river in the town. Located on the river's east bank, immediately below the lower falls, the Webber-Hawthorne structure measured thirty feet by forty feet and was built using locally harvested burr oak timbers with a foundation made from heavy quartzite slabs.

The formal opening of the business was described as the biggest event of the year. Finally, wrote one observer, "the water that for so many centuries had been expending its energies in simply wearing a channel through the rocks, should be harnessed and controlled by the inventive genius of man." Cascading water had long been a soundtrack for the town. Suddenly, there were new noises at the falls of the Big Sioux.

Harnessing water at the falls was viewed as a key to the city's growth. A local newspaper, the *Pantagraph*, touted the benefits of water-powered mills. The falls, gushed the newspaper, "are the cradle of that great hydraulic wizard whose wand is to turn our wheat into flour…and make us a manufacturing metropolis."

Four years after Webber-Hawthorne opened for business, construction began on a larger milling operation. The Cascade Mill included erection of a substantial dam across the river channel, above the power plant. Measuring 190 feet long and 16 feet high, the dam utilized three water release gates, each 5 feet by 6 feet. The structure's indestructible composition was capable of withstanding ice flows and surging floods. The 16-foot-wide base, composed of closely spaced vertical iron rods, stones and concrete, was fastened to the river's floor within a 2-foot groove carved into quartzite bedrock. This dam could not be undercut. Above the dam's substantial masonry were heavy pine planks fitted onto the exposed top portion of the iron rods. The boards added 12 inches to the height of the dam and were detachable. Five hundred loads of broken rock protected the dam's upriver side. The reservoir backed up behind the dam lifted water levels within the river's slender, twisting channel for three miles upstream. The *Pantagraph* described the dam as "the most substantial in all this country," adding that the "power it will permanently afford is one of the factors in the future of our town." Opening a single gate, said the newspaper, could power the Cascade mill, leaving the other two to furnish power to other manufacturing enterprises.

Using that dam to control and direct water flow added reliability to the mill's waterpower. There is no way to know the full impact of the Cascades Mill dam on the river's hydrology and the falls, but it must have been significant. The milling enterprise produced a higher-grade flour than its Big Sioux River rival. And it possessed a larger capacity—between one hundred and two hundred bushels per day—that could serve more farmers in a wider area and produce far more flour than area residents needed.

By the time Sioux Falls' first water-powered facilities were operational, technological improvements in energy production were remaking water mill design. The traditional vertical waterwheel was being replaced by efficient metal turbines. The vertical waterwheel had been a popular device, and modern imaginations remain fixated on this design as a streamside icon from long ago. But it is likely turbines were favored over waterwheels at most power plants along the river and cascades in Sioux Falls.

By 1878, Sioux Falls enjoyed railroad service, allowing harvested crops to be shipped easily from area fields to Cascades Mill and finished flour to penetrate markets outside the Sioux Falls region. The city's population had reached 2,000, and farm sites in the vicinity were being claimed at a breathtaking pace. In one month during 1878, the federal land office in Sioux Falls released 200,000 acres of rural land to qualified settlers. Land office staff were overwhelmed, processing as many as 134 claims in a single day. During the next dozen years, 8,000 new residents arrived in Sioux Falls—and the town's rapid growth wasn't a geographic anomaly. The so-called Great Dakota Boom, from 1870 to 1890, saw the population of what is now South Dakota increase from less than 12,000 to more than 325,000. Many of the newcomers were immigrants. Not surprisingly, the number of farms and towns grew at a staggering rate. Farms in 1870 numbered 1,700. Twenty years later, there were 50,100 farms. The number of platted towns increased from 6 to 310.

Quarrying Sioux Quartzite became a vital industry in Sioux Falls that would not have been possible without the railroads. Indeed, the first cargo shipped by rail from Sioux Falls was blocks of quartzite building stone. Several large quarries were developed in the city's early years, and while most of the city's quarries were in eastern Sioux Falls, some activity occurred along the cascades.

The market for quartzite expanded to a variety of uses. In 1880, a delegation of Sioux Falls business leaders stood before the Omaha City Council to promote Sioux Quartzite as a material to pave city streets. The council debated the merits of using granite versus the suitability of quartzite. Two professors from Nebraska's state university claimed granite was superior. But the Sioux Falls contingent presented the science of pioneering geologist C.A. White. Sioux Quartzite, said White in 1867, is "absolutely indestructible." The clincher was a written endorsement from Chicago's street superintendent, who said Sioux Quartzite was the best stone he'd used for paving. The city council approved purchasing and utilizing Sioux Quartzite to strengthen and overlay Omaha's streets that day. A lucrative new market for Sioux Falls' native stone had been secured.

The railroads sought more flour and wheat as freight. The success of the first two flour mills prompted development of the grandest mill of all. Named the Queen Bee, this facility became a symbol of Sioux Falls' rising aspirations. When construction started in August 1879, one observer said the Queen Bee mill would be "the finest flouring mill on the continent." Another proclaimed the mill was the most ambitious waterpower project in the entire Northwest.

A big project like the Queen Bee was the invention of a big dreamer. That man was Richard Pettigrew, a fellow who would, in a multitude of ways, leave an indelible mark on Sioux Falls and South Dakota. Pettigrew, a Vermont native, came west in 1869, after studying law at the University of Wisconsin. Not yet twenty years old, he took a temporary job as a chainman on a government survey crew headed to the western frontier. When he arrived, practically penniless, in Sioux Falls, little more than a military installation, Fort Dakota, existed there. Pettigrew proved proficient and responsible, and he was promoted to lead the survey team. After the group assessed lands along the Big Sioux River, Pettigrew acquired property in Sioux Falls. He then returned to Wisconsin to complete his legal studies before traveling back to Sioux Falls. Along the way, he stopped in Sioux City to purchase the materials he needed to construct a building for his soon-to-open law practice and real estate business. Using a horse-drawn wagon, Pettigrew hauled the materials to his new hometown in 1870.

Pettigrew continued his work as a surveyor and secured government contracts in 1872 and 1873 to determine township boundaries in present-day North Dakota, between the Red and Missouri Rivers. On one surveyor's mission, he hunted buffalo from horseback with Sioux Indians near the Canadian border, describing his companions as "wonderful people." Not only could Pettigrew calculate meridians and standard parallels, but he also possessed the skills of a frontiersman and the shrewd guile of a politician. It was Pettigrew who had thwarted Boston speculators' attempts to acquire land near the falls when Fort Dakota was shuttered.

Pettigrew disclosed that he chose Sioux Falls over Yankton to be his new home because, longitudinally, Sioux Falls sat on top of a string of cities stretching from Texas to Kansas City, Omaha and Sioux City. Evidently, his work as a surveyor emphasizing straight lines and where they pointed made an impression on him. Pettigrew praised the fertility of land in the Sioux Falls region and was convinced that the fledgling town could fulfill its promise by attracting more industrialization and railroads. He became an avid promoter of Sioux Quartzite as a construction stone.

As early as 1873, Pettigrew pitched Sioux Falls as a rail destination, eventually convincing the owners of the St. Paul and Sioux City Railroad to build a branch line to Sioux Falls, which was completed in 1878. It was the first of five different railroads that would extend to Sioux Falls in those early years, making the community a genuine rail center. One of those railroads, the Rock Island Line, agreed to serve Sioux Falls after Pettigrew made numerous self-financed trips to Iowa, Chicago and New York.

Pettigrew's service in the territorial and state legislatures showed him to be smart, stubborn and feisty. At least once during a lawmaking session, he engaged in a violent fistfight, a fracas involving the territory's secretary, a large, imposing man who had served as a general during the Civil War. Pettigrew gleefully reported that he'd whipped the bully. Pettigrew's first election to the territorial legislature was clouded by accusations of shady ballot manipulations, but his accomplishments for Sioux Falls were significant, including snatching the territorial penitentiary from the community of Bon Homme. Later, as South Dakota's first U.S. senator, he vigorously supported women's suffrage and convinced federal officials to construct government buildings in Sioux Falls using locally quarried quartzite. He also persuaded his friend Andrew Carnegie to build a public library in Sioux Falls—to be constructed of Sioux Quartzite, of course.

Wilmot Brookings and his pal Josiah Phillips owned prime parcels at the falls, but neither created businesses to use the waterpower available to those parcels. They were sitting on their land, it seemed, waiting for a developer with deeper pockets to come along. Pettigrew found the well-to-do outsiders who would buy and develop Brookings's riverside parcel, including E.F. Drake, a St. Paul–based industrialist, and George Seney, a wealthy New York banker. It was Drake who built the first rail service to Sioux Falls. Seney invested the lion's share of the half a million dollars needed to develop the new Queen Bee milling project. No ordinary structure, the Queen Bee would be built entirely of Sioux Quartzite.

Pettigrew had likely met George Seney because the New Yorker and his bank financed and owned railroad projects around the country. In 1882, Seney sold one of those rail lines, the Nickel Plate, for $7.2 million in gold.

In a lavish New York City office where priceless masterpiece paintings hung on the walls, George Seney must have carefully studied the man sitting across his desk asking him for a small fortune. The fellow from the western frontier was well-dressed and knew how to converse in a language Seney appreciated. It was evident, noted Seney, that Richard Pettigrew chased enormous projects with indomitable swagger, an admirable trait. Though Seney wasn't familiar with Pettigrew the buffalo hunter, wilderness surveyor or wide-eyed pioneer who, not many years earlier, hauled construction supplies by wagon to build his own business headquarters in a raw, undeveloped village named Sioux Falls, he understood that the man seeking investors for a project on a distant river lived a life requiring a different sort of resourcefulness and resilience than he or his urbane business partners possessed. *Pettigrew moves comfortably in my world,* the refined financier might have reflected, *but could I move comfortably in his?*

Richard Pettigrew founded the Queen Bee mill. No early citizen had a greater impact on Sioux Falls and the cascades of the Big Sioux River. *Library of Congress.*

George Seney purchased land near the falls for $45,000 and then formed a new corporation, the Sioux Falls Water Power Company, to build and manage the new mill. As part of the property transaction, Seney acquired the wooded island immediately upriver from the falls, a place locals called Brookings Island. This special location would soon assume a different name: Seney Island.

At the time, there were no laws or any type of enforceable regulations restraining or moderating what could be built at the falls—or along any river. There was no such thing as a water right, and there were no policies intended to limit water withdrawals or use. If you owned property along the river, you could impose your will on the river. Only a shortage of money could prevent a shoreline property owner from building a dam, weir or intake system to serve a water-driven power plant.

Circulating in Sioux Falls was a rumor that Pettigrew had duped Seney into supporting the milling enterprise. It's a witty story, but those who study Pettigrew conclude that it is a fable, fueled, perhaps, by Pettigrew's awkward place in Sioux Falls lore. Although his reputation as a devoted city patron was deserved, his confrontational tactics in politics and business earned him many enemies.

While raising money for the Queen Bee, declares the tale, Pettigrew invited Seney to Sioux Falls to visit the site. River flows in the days leading up to Seney's arrival were lower than Pettigrew liked, so he had a small dam constructed upriver from the falls. At a strategic moment before Seney reached the falls and stepped from his carriage, Pettigrew signaled his men to open the temporary barrier. As the New York banker stood beside the river, an artificially elevated surge of water poured past. "Most impressive!" Seney may have exclaimed. And he decided to invest.

The allegation is easy to believe, especially if you're predisposed to disliking Pettigrew. However, Pettigrew historian Wayne Fanebust found no evidence that George Seney traveled to Sioux Falls before construction of the Queen Bee. It is possible the myth had its roots in a genuine deception orchestrated by Pettigrew. Sometime during the warm months of 1878 or 1879, before the Queen Bee plan was launched, Pettigrew and Sioux Falls hosted a trainload of eastern newspaper editors and journalists. The visitors desired a firsthand look at a thriving, remote town in the West and were most interested in describing the area's economic prospects for their readers. This was precisely the type of promotion Pettigrew wanted. But as he prepared for the important callers, he noticed that the river was trickling over the falls. "I knew," he later wrote, "that our much-advertised falls and water power would be a terrible disappointment to our visitors." And that meant, he anticipated, that the stories generated by the newspapermen would "do Sioux Falls a great injury."

A short distance upriver from the falls was a dam spanning the river, where captive flows trickled over the top of the structure. It is uncertain whether the dam allowing Pettigrew's ruse was the Cascades Mill dam or some other dam, but a crew was hired to install temporary planks to add height to the barrier. "This maneuver," explained Pettigrew, "raised the river above the dam twelve inches." He admitted that just before the newspapermen arrived, he ordered his men to remove the planks, and pooled water rushed downriver and over the falls. According to Pettigrew, "Our visitors were delighted and newspapers in the South and East were filled for weeks with descriptions of our beautiful falls and great waterpower, equal to that of Minneapolis, almost equal to Niagara, so various accounts read."

Some years earlier, when the falls ran high and impressive, Pettigrew had commissioned a photographer from St. Paul to capture flattering stereopticon images of the scene. During the journalists' visit, Pettigrew gave each of them a choice photo from that earlier shoot, and he was delighted to later learn that accompanying many of their newspaper articles extolling Sioux Falls was a photo he'd provided. "The advertising given Sioux Falls was of great value and started a boom which brought many people and much money to our city," Pettigrew claimed. "It was a start in building our city."

When Queen Bee construction began, more than five hundred people gathered to honor Richard Pettigrew's magnificent plan. Dignitaries, investors and community residents were on hand to celebrate. The crowd retreated to a nearby blufftop overlooking the site before the first charges of dynamite were detonated on the river's shoreline to crack and clear bedrock so the mill's foundation could be positioned. Engineers also intended to blast rock from the falls area to provide the material necessary to construct the mill and manipulate the channel.

Only months later, construction of the Queen Bee was interrupted by a powerful flood that devastated poorly protected, poorly planned Sioux Falls. Snowfall amounts in the Big Sioux watershed during the winter of 1880–81 were the greatest on record, and Sioux Falls became isolated from the rest of the world, with all forms of inbound and outbound travel suspended. The community's 2,100 residents lacked many ordinary provisions as winter wore on, and no one, not even a railroad, could break through the massive snowdrifts. Marooned, Sioux Falls citizens faced the cold without wood or coal, prompting the Worthington and Sioux Falls Railroad Company to present the public with thousands of rail ties stored at their Sioux Falls yard. These heavy timbers could be cut, split and burned for warmth. A generous discount made the improvised fuel affordable for all.

A sudden thaw during the spring of 1881 delivered enormous runoff down the valley. Flows in many tributaries and the Big Sioux itself were blocked by formidable ice barriers, causing immense pools to form upstream from the ice. When those frozen flows finally broke apart, torrents surged downstream, ramming hefty ice chunks ashore, where they smashed into trees, bridges and buildings. High flows swarmed across floodplains throughout the river valley. On April 20, in Sioux Falls, at the falls, the Webber-Hawthorne mill was ripped from its foundation and floated a short distance downstream.

Lone Rock, located in the lower cascades, was a popular landmark before being toppled by the ferocious 1881 flood. *Courtesy of Robert Kolbe Dakota Collection.*

A large metal pipe running along the river to serve the unopened Queen Bee mill was torn from shore and washed away. Numerous low-head dams had been built across the river, above and below the falls, and all were destroyed. The only intact survivor was the stout, reinforced structure at the Cascades mill. A Sioux City newspaper reporter described the 1881 Sioux Falls flood as a rebellion of the river: "The waters that had for so long acted as a servant, took the position of master."

One journalist stationed on high ground wrote that buildings were bobbing past faster than he could identify them. A blacksmith's shop floated by, followed by a saloon. The wares of local lumberyards—boards, posts, planks, shingles and lath—swept past. Then a lumber store. Next came a cigar factory. Then a restaurant, three bridges, office buildings, a livery stable and a sash factory. Like big buoys, the buildings or their remnants came to rest in shallow water at the edges of the flooded channel. Nine years after angry Sioux warriors burned down nearly all of the town, water and ice submerged and destroyed much of what had been rebuilt.

A beloved quartzite landmark, rising up like a small, elevated island in the midst of the falls, the so-called Lone Rock, was sheared from the channel floor by fast water and heavy ice and cast into a twenty-foot-deep hole downstream, where it was inundated and invisible. It is likely a large fracture existed near the base of Lone Rock, weakening the formation. Before the flood, when the river was at average depth, the flat-topped eight-foot-by-eight-foot pinnacle perched a sitting man four feet above the waterline. The prospect of watching the falls from that intimate vantage attracted many boat-borne sightseers, and anglers caught blue ribbon pickerel from there.

What had already been completed of the Queen Bee mill withstood the flood, though its wooden office building drifted away. A dam to serve the mill was salvaged. As flood flows subsided, construction on the mill resumed. Miraculously, on October 24, 1881, only seven months after much of Sioux Falls was swamped, the Queen Bee began grinding locally grown wheat.

Top hats may have been tossed high as the finished mill was opened and the sound of machinery rose above the murmur of cascading water. Sioux Falls' population had grown to 2,800. Wheat production in the immediate region was estimated at 250,000 bushels per year. Community optimism was palpable and pervasive. So monumental and symbolic was the Queen Bee that a popular nickname for Sioux Falls began to circulate: the Queen City.

Richard Pettigrew hadn't exaggerated when he told George Seney the Queen Bee would be among the most impressive mills in the West. Considered only for its size, the structure was startling. The upcoming generation of American skyscrapers, starting in 1883, added buildings twelve stories high or higher to the skylines of Chicago and New York. The Queen Bee soared seven stories and was hailed as the tallest building between the Mississippi River and the Rocky Mountains. The structure's footprint was nearly as impressive as its 104-foot height. Measuring 80 feet by 100 feet, the milling plant seemed vast, limitless.

Top: Looking east across the cascades to the future site of the Queen Bee mill. *Courtesy of Robert Kolbe Dakota Collection.*

Bottom: The massive Queen Bee mill, opened in 1881, industrialized the river's shoreline and disfigured a section of the falls. *Courtesy of Siouxland Heritage Museums.*

Other structures associated with the mill, including administrative offices, a grain storage unit and a four-story elevator/warehouse for finished flour, were part of a complex that crowded the cascades. On-site storage capacity for wheat measured at least 100,000 bushels. Daily flour production could reach 1,200 barrels. Connecting the mill to the city's rail lines was a siding accommodating twenty-two freight cars.

The milling facility was situated slightly set back from the river, and an expansive plumbing system steered water from the upper falls and channel to the power plant. That conveyance utilized a series of channel revisions to harness the natural river and rework the riverbank. A lengthy, brawny dam, almost six hundred feet long and ten feet high, connected the river's eastern shoreline to Seney Island, and a weir connected the western side of the island to the river's western shore. Engineers specified the dam be built diagonally rather than straight across the river. The angled design, said the engineers, would preserve or even enhance the appearance of the falls. This claim was either misleading or mistaken. The dam's diagonal direction actually served a more utilitarian objective. Flows were collected in the millpond behind the dam and deflected by gravity across the face of the dam and the width of the river. On reaching the eastern shoreline, these directed, impounded flows were funneled into an intake system and squeezed into a gatehouse where a large, heavy plate, the gate, could be opened or closed to control flows entering a spacious iron pipe, seven feet in diameter. The pipe, cradled by masonry piers and visible as it ran along the river's edge for several hundred feet, followed the contours of the shoreline and the falls. Dropping at a severe angle, piped water plunged downward, gathering speed and force before entering the power pit of the turbine house and forcefully compelling a hydraulic turbine spinning thousands of revolutions per minute. That turbine was attached to a vertical driveshaft that powered the mill's main shaft, located above the turbine house and within the mill. Richard Pettigrew claimed the system could produce up to one thousand horsepower, an exceptional engineering achievement.

Inside the flour factory was an array of notable features. One historian, Dana Bailey, declared that the materials and workmanship used to build the mill exceeded those found in any other manufacturing structure in the Northwest. "Nothing but the latest and most improved machinery was purchased to equip this remarkable mill," he wrote. To accomplish the mill's purposes were ten miles of belting, three miles of conveyors and two miles of elevators.

Visible is the large iron pipe that fed fast-moving flows to the Queen Bee turbine house. *Courtesy of Ed Monson Collection.*

Visitors to the falls held their breath as they gazed upward at the towering Queen Bee structure. The new mill replaced the natural cascades as the chief aspect of this long-honored spot. Why look downward at mere water and rock when a monumental stone edifice lorded above all else? The value and purpose of the cascades were dramatically changed. Where once there had been a natural shoreline, there stood a warren of industrial structures. The sound of water moving over rock at the main section of falls was suppressed by the sound of industry.

Sioux Falls had officially joined another milling giant as a hub of agricultural processing. Charles Pillsbury opened the largest flour mill in the world at St. Anthony Falls on the Mississippi River in Minneapolis the same year the Queen Bee began grinding wheat. Pillsbury's colossus was powered by two fifty-five-inch turbines, each generating 1,200 horsepower, that could produce four thousand barrels of flour each day. Minneapolis became the undisputed center of the milling universe. Sioux Falls business leaders wanted their community to likewise achieve a notable level of respectability as a value-added agriculture hub. And for several years, that promise was fulfilled. So much flour was shipped in barrels from the mill during its first year of operation that forty coopers (the tradesmen who built wooden barrels from scratch) worked steadily on-site at the facility.

Unfortunately, the elation didn't last long. Within two years, the ballyhooed milling operation floundered and sank into bankruptcy. Inadequate, unpredictable waterpower related to fluctuating and seasonal river flows was cited as a central reason for this spectacular failure.

Richard Pettigrew and other like-minded capitalists fancied themselves as shrewd, calculating investors. But their understanding of the Big Sioux River, a small river flowing through semiarid prairie, was flawed. Perhaps these speculators were guilty of willful misunderstanding or were blinded by optimism. The same falls that astonished Moses Armstrong bored Philander Prescott. Covered by snow and flowing low, the falls viewed by Prescott were completely different than the falls heard by Armstrong from three miles away.

Sioux Falls pioneers had long compared the falls on the Big Sioux River to the falls in Minneapolis on the Mississippi River. This was fantasy. Above the falls in Minneapolis was a vast, moist watershed feeding a growing, steadily flowing Mississippi. The natural cascades at Sioux Falls were certainly dramatic and scenic, especially when Big Sioux flows were high during snow thaw or after heavy rains. But the calendar wasn't kind to waterpower at Sioux Falls. By autumn, practically every year, the river's capacity to energize industry dwindled.

The Queen Bee postmortem should have identified the river's industrial liabilities, but there was no stopping wide-eyed speculation. Richard Pettigrew was unfazed and unrelenting, a fountainhead of confidence. He looked at the falls and the river, and all he saw was untapped manufacturing potential that would be the linchpin in Sioux Falls' growth.

In 1883, former Queen Bee manager James Drake, nephew of E.F. Drake, opened an unusual niche business at the falls. Drake Polishing Works used moving water to power polishing beds that buffed and shined quartzite extracted from Drake's nearby quarry. Drake also owned quarries near St. Cloud, Minnesota, where so-called red granite was procured and shipped to his Sioux Falls polishing works. The buffing process employed by Drake

Channel modifications and dam downstream from the Queen Bee mill. *Courtesy of Robert Kolbe Dakota Collection.*

created a dazzling, lustrous building stone that was fashioned into columns and other forms of ornamentation used to accent the exteriors of fine homes and buildings across the country. Gravestones and monuments were another staple of the business. Despite reaching eighty employees, the company couldn't keep up with orders. Early promotions described Drake's enterprise as the nation's largest rock-polishing business.

Drake was inventively entrepreneurial. He added a rare Belgian stone, petrified wood from Arizona and chalcedony to his line of polished stone products. High-end retailers like Tiffany's of New York wishing to create one-of-a-kind furniture, lamps, jewelry, clock faces and other specialty items became clients. A small tabletop, measuring just thirty inches in diameter, sold for $1,100. The nation's wealthiest people sought decorative treasures that originated at Drake's factory.

Drake's stone artistry gained worldwide attention when the company displayed its products to an international audience at the 1893 World's Columbian Exposition in Chicago. But his businesses were hurting by the late 1890s as the quarrying industry struggled. Expanding uses of improved concrete narrowed commercial options for Sioux Quartzite, especially as a street paver. Drake lost his Sioux Falls quarries, and production at his polishing plant began to shrink. His primary source of petrified wood, a specific area near Prescott, Arizona, was awarded protected status by the

James Drake's rock-polishing business used cascades waterpower and attracted a national clientele. *Courtesy of Siouxland Heritage Museums.*

government, prohibiting collection of petrified wood there. Downsizing extended the life of the business, but by 1905, the plant had closed, and Richard Pettigrew bought Drake's unsold stock of polished petrified wood. Pettigrew later used some of his acquisition to embellish a large addition he built for his stately residence. Fifty-three pieces of polished petrified wood were used as ornaments beautifying an arch and sign welcoming visitors to a large Sioux Falls cemetery named Woodlawn that Pettigrew started on property he owned. The nation's Museum of Natural History in Washington, D.C., asked Pettigrew to supply several sections of petrified logs. "I sent them by express," said Pettigrew.

Quarrying activities were not uncommon near the falls. Prisoners from the territorial penitentiary quarried quartzite there and hauled it up a nearby bluff to construct buildings and a wall around their prison. Numerous commercial quarriers mined the rock. City-owned quarrying operations also excavated quartzite.

A locally owned railroad outfit named the South Dakota Central desired to build a line northward, up the Big Sioux River, linking Sioux Falls to Watertown and serving numerous smaller farming communities in between. A real estate venture associated with the railroad, the South Dakota Land

Company, peddled and promoted property along the rail route, particularly in new and existing trackside towns. A resort was proposed for Lake Poinsett, north of Brookings, and summertime transport for tourists visiting this play place was a component of the South Dakota Central's business plan.

Founded in 1904, the South Dakota Central attracted widespread support in Sioux Falls. But the railroad's bigger, older rail siblings were not pleased with the upstart's ambitions. The Milwaukee railroad company tried but failed in federal court to block Central's proposed route.

The South Dakota Central, based out of the Illinois Central depot in east-central Sioux Falls, needed to cross the Big Sioux River at some point in order to travel north. This would be accomplished, declared the railroad company, by building a single-track bridge over the river at the falls.

Opposition to this location for a significant river crossing was subdued, if any existed at all. Sioux Falls' namesake and shrine to stunning natural scenery would be further blighted by a bridge and trains noisily passing overhead. It appears there were no permits required to build the bridge.

Right-of-way land acquisition for the rail line's route was in full swing during the autumn of 1905, and condemnation proceedings, exercised by the railway, were part of that plan. The president of the South Dakota Central journeyed to Chicago, and on December 11, 1905, he closed the deal to finance the bridge. Construction commenced just eighteen days later. No glitches were reported, and work proceeded quickly. On February 28, 1906, to celebrate completion of the structure, a ceremonial train traveling from the north crossed over the falls. There was a triumphant whistle as the train pulled into its station a short distance south of the new bridge.

By 1884, the same year James Drake opened his rock-polishing enterprise and not long after the Queen Bee closed down, industrialization on the Big Sioux River near the falls included a modest hydroelectric facility named the Electric Light Company that supplied electricity to Sioux Falls' streetlight system. At the time, the city's downtown was illuminated at night by both gas and electric lights.

Electricity was gaining popularity across the nation as science improved technology and equipment. Hydropower first emerged as a source for electricity in 1850, and a commercial facility providing power to light a factory in Grand Rapids, Michigan, opened in 1880. The following year, a mill in Niagara Falls, New York, began producing electricity to power streetlights in that city. Four years later, the world's largest hydropower plant was operational there. In 1882, Thomas Edison's Pearl Street Station in New York City used coal to generate direct current electricity and power

Top: Originally built in 1906, this railroad bridge delivered pollution, noise and a diminished aesthetic to the cascades. *Courtesy of Ed Monson Collection.*

Bottom: The Queen Bee mill, circa 1907, across the river from land that would later host the first components of Falls Park. The new 1906 rail bridge is also shown. *Courtesy of Ed Monson Collection.*

more than one thousand light bulbs in homes and businesses surrounding the plant. By 1900, there were hundreds of small hydropower plants operating across the country, powering factories, businesses and appliances and lighting in homes. The availability of newfangled electricity doomed water-powered gristmilling.

The owners of the Cascades Mill diversified their business portfolio by purchasing the Electric Light Company in 1887. New investors upgraded the electric plant, including housing electrical equipment in a stone building near the mill, a short distance upriver from the falls. The plant added steam-generated electricity to hydro production and generated direct current electricity that traveled a short distance by transmission line to streetlights in the commercial district and to other users.

In 1901, the City of Sioux Falls developed a municipal utility to serve a portion of the community's streetlights. Across the country, the benefits and disadvantages of publicly owned utilities providing electricity or water were debated. Some described public utilities as undesirable socialism. Sioux Falls' municipal power company dented profits at the Cascades Milling Company, but there was plenty of electricity business to share. As electricity gained popularity and as Sioux Falls grew, so too did the fortunes of Cascades Milling Company. Light bulbs were suddenly everywhere, in many rooms in many buildings. When the hydro-steam plant started operations, approximately four hundred light bulbs were fed electricity by the facility. Within several years, that number rose to five thousand.

Until 1905, private sector electric generation serving Sioux Falls was dominated by Cascades Milling Company. Competition appeared when a new company, Sioux Falls Electric Light and Power, was incorporated and sited a generating plant alongside the lower falls, at the former location of the Webber-Hawthorne mill and Drake's rock-polishing factory. This plant used Drake's wheel pit to produce hydropower and also burned manure purchased at local livery stables to create steam-powered electricity. Soon, the rival companies were fighting over contracts and opportunities.

Closely watching the instability and conflict plaguing electricity generators in Sioux Falls was Henry Byllesby, a utility entrepreneur based in Chicago. His favored business strategy was to identify promising communities lacking well-organized or well-financed electric utilities. Then he'd swoop in, purchase the competition and develop a bigger, more effective utility.

Henry Byllesby wasn't some fly-by-night dabbler. He'd learned the technical, commercial and industrial aspects of electrical generation and sales while working from 1882 to 1890 for electricity pioneers Thomas Edison and George Westinghouse Jr. Although he'd dropped out of the engineering program at Lehigh University, Byllesby invented and oversaw the design of more than forty patents for electric lighting devices while working at Westinghouse. But his real interest was business, not science.

Before his thirtieth birthday, Byllesby had run the Canadian division of Edison's company, and he later helped incorporate Westinghouse Electric, serving as general manager of the new company in the dizzying and momentous early days of an industry destined to change the nation and the world. His remarkable climb continued when he resigned the Westinghouse position to assume the presidency of Northwest Thomson-Houston Electric Company, a large electrical systems company based in St. Paul. The nation was discovering the value of electricity, and Edison, Westinghouse, Thomson-Houston and others were vying for contracts to build electrical systems, generating stations, transformer stations, distribution systems and electric rail and lighting networks. A critical issue of that era was to determine whether alternating current or direct current systems were superior. Thomas Edison was a proponent of direct current but later admitted he was wrong. In 1891, the so-called Father of Electricity tried to acquire Thomson-Houston and its alternating current patents but failed. Instead, Edison's company was purchased by New York banker J.P. Morgan and merged with Thomson-Houston to form a new company called General Electric.

Byllesby became vice president of the Portland (Oregon) General Electric Company, where he designed and built four hydroelectric facilities in four years. He then moved to Chicago in 1902 to establish his own utility development business. Buoyed by wealthy investors, Byllesby's company began to buy, manage and expand utility companies across the Midwest.

In Sioux Falls, Byllesby and his team acquired the Cascades power plant and the onshore buildings, property and in-channel infrastructure that had once served the Queen Bee mill. The newest company operating along the falls, Byllesby's company, would be named Sioux Falls Light and Power.

A short distance north of the Queen Bee mill, on the east side of the falls, Byllesby intended to site a handsome operations center. The upper portion of the facility, built of Sioux Quartzite, would look down at the river from an imposing perch and would be anchored to the shoreline by a twenty-foot-tall concrete foundation.

Utility magnate Henry Byllesby built this powerhouse along the cascades and altered the falls to serve his business. The Queen Bee mill is also shown. *Courtesy of Ed Monson Collection.*

Dry channel downstream after Henry Byllesby acquired and refurbished the Queen Bee dam. *Courtesy of Siouxland Heritage Museums.*

To compensate for uneven, often inadequate Big Sioux River flows, Byllesby enhanced river control features and impoundment capacity. Queen Bee's original dam, 590 feet long, more than 10 feet tall and still blocking the river, was shored up and connected to a new retaining wall following the river's western shoreline halfway to Sixth Street. The river's eastern bank was also stabilized by concrete and riprap stretching hundreds of feet upriver. The dam would pass beneath the new South Dakota Central bridge, and when river flows rose to the top of the dam, the millpond behind the dam would reach within 3 feet of the crossing's deck.

The dam and retaining wall would prevent the river from supplying flows to the secondary channel that bordered Seney Island's western shore. No longer would Seney be a genuine island. These "improvements" would both block and pinch the channel, creating artificially pooled flows that would, declared Byllesby officials, offer Sioux Falls a source of boating recreation. Apparently, they felt that rowing a boat up and down a thin ribbon of reservoir bordered by factories and warehouses would provide a charming outdoor experience.

Above: Concrete and rock walls choked the river channel and cascades to harness water for industry. *Courtesy of Ed Monson Collection.*

Opposite, top: Channel and cascade destruction by Henry Byllesby's utility. *Courtesy of Robert Kolbe Dakota Collection.*

Opposite, bottom: Crews working for Henry Byllesby destroyed sections of the cascades and carved an artificial channel to enhance utility performance. *Courtesy of Siouxland Heritage Museums.*

Immediately downriver from the millpond, starting at the face of the dam, the river channel would be impacted in a completely different way. Instead of being permanently flooded by a reservoir, the channel would be deprived of water. One journalist shared what he learned from a Byllesby engineer. During periods of low or ordinary flows, said the engineer, all the river's water would bypass the most dramatic stretch of the falls. In other words, for at least some of the year, part of the cascades would be dewatered, reducing the spectacular falls to dry stony clefts and ledges.

Nearly as important as directing flows *to* the power plant to generate electricity was the process of moving water *away* from the plant. To facilitate this, a tailrace was created when Byllesby's engineers enlarged channel capacity immediately downriver from the power plant. Quartzite formations in the channel, comprising what had been known as the lower falls, were obliterated by dynamite and steam drills. A natural channel became an artificial channel. Rock fragments and rubble were

then hauled away or stacked as riprap. G.G. Caldwell, superintendent of construction for H.M. Byllesby & Company, would later describe the impact of his project: "I've destroyed the most scenic place in Sioux Falls," admitted Caldwell.

Inside Byllesby's operations center and power plant was the latest technology. Three turbines would have the potential under optimal river flows to create three thousand kilowatts of electricity. Extra space existed to add a fourth turbine. A network of transmission lines extending fifteen miles from the plant would deliver power to customers in the city and the countryside. The company's first contract in Sioux Falls was to supply the electricity necessary to operate the community's streetcar system.

"The [electrical] power which can be furnished by the water of the Big Sioux River is almost limitless," exclaimed the Sioux Falls *Argus Leader* in May 1908. The scale and impact of this great new industry for both Sioux Falls and the entire state of South Dakota, stated the newspaper, rivaled the discovery of gold in the Black Hills.

Six months later, the *Argus Leader* celebrated Light and Power's "victory" over the falls. "The removal of the classic Sioux river from the realm of inspiration for poetry and the placing of it in the list of commercial assets of the city of Sioux Falls and the state of South Dakota is an accomplished fact," the newspaper reported.

Byllesby's next move was to buy out the group running Sioux Falls Electric Power and Light. His takeover at the falls was now complete. Soon, Byllesby had a fleet of electricians installing electrical service inside homes and businesses. From a downtown storefront, the company peddled residential appliances, including irons, fans and washing machines.

In 1909, Byllesby reorganized his many electric utilities into Consumers Power Company, and he purchased more utilities in North Dakota, South Dakota and Minnesota. Seven years later, he renamed his flagship company Northern States Power (NSP).

Byllesby sold the Queen Bee to Minneapolis-based United Flour Milling Company in 1911, and the new owners announced the mill would be reopened as a world-class milling facility powered not by water but by electricity.

In 1913, Byllesby built a modern energy plant containing coal-powered steam boilers alongside the original generating facility. The company needed to diversify its electrical supply to overcome power shortages occurring when low flows through the falls were unable to reliably drive turbines and power dynamos. Other features were added to the company's cluttered industrial

Northern States Power downstream channel impacts, including the tail race infrastructure, wiped out a stretch of the lower cascades. Taken in 1934. *Courtesy of Siouxland Heritage Museums.*

Henry Byellsby's utility transformed a stretch of river and cascades into a deep pool and trench. *Courtesy of Siouxland Heritage Museums.*

Sioux Falls Light and Power upstream infrastructure at the cascades included a hefty dam, concrete channel constrictions and a gated water intake. *Courtesy of Ed Monson Collection.*

Sioux Falls Light and Power became Northern States Power, and the company's industrial footprint at the cascades was expanded. *Courtesy of Ed Monson Collection.*

campus overlooking the falls, including a water tower, coal cranes, coal storage structures and smokestacks, one of which towered 220 feet.

Within a decade, power generation at the Sioux Falls plant reached eight thousand kilowatts. The Illinois Central railway had long been an important connection for Sioux Falls to the coal fields of southern Illinois, and Henry Byllesby relied on that coal for a portion of his Sioux Falls production. But in 1933, a natural gas pipeline reached Sioux Falls, and Northern States Power switched its facility to gas and fuel oil. This was Northern State Power's first such conversion among the many plants in its system, and hydropower faded as a source of electricity in Sioux Falls. To accomplish power generation, permanent physical change had been inflicted on the river and the falls, but the commercial and social benefits resulting from those modifications lasted only a brief time.

Chapter 6
THE SAGA OF SENEY ISLAND

Sioux Falls was born on Seney Island in 1856 when two representatives from the Western Town Company, Ezra Millard and David Mills, built a small, simple dwelling there.

Field notes produced during an 1857 survey by the federal government's General Land Office provided a dutiful description of the island: "There is an island in the river just above the falls containing about 12 acres with good timber on it—oak, elm, ash and linden, with some cedar among the rocks at the falls." That was the exact entry: no color, no hype, but enough information to indicate that the island was sizeable and timbered.

Others applied more colorful language. Although limitless homesites existed in the vicinity, newcomers Mills and Millard chose the island because it was an irresistible setting. "It was just above the place where the waters of the Sioux plunged over a series of rock ledges," wrote historian and author Charles Smith. "The grandeur of this view was indescribable."

Words like *glorious* and *spectacular* were among the superlatives often used to portray this forested isle at the crest of a showy set of falls. Jacob Ferris, the author whose flattering descriptions first drew town builders to Sioux Falls, included the island as one of the townsite's special attractions. George Staples, founder of the Western Town Company, recognized the island's value as a haven where residents and visitors could gather together to celebrate nature and fellowship. "Above the falls," Staples wrote in a report to company shareholders, "the river is divided and forms an island…which is covered in dense timber—a fit place for pic-nics, Fourth of July orations,

Across the cascades and the river's main channel is Seney Island's wooded and rocky shoreline. *Courtesy of Siouxland Heritage Museums.*

and political gatherings. So soon as the river unites below the island, the waters begin to tumble over the successive strata of rock as they crop out."

Staples's prediction that the island would become a cherished meeting place and nature sanctuary was soon fulfilled. People living in early Sioux Falls never tired of the remarkable scenery. Out-of-towners, particularly farmers and their families residing on treeless prairie, flocked to the island, drawn by the cool shade of mature hardwoods. Along the island's western channel, the forest canopy hung low, almost dipping into the water. In the center of the island was a grassy meadow colored by wild rosebushes and other native flowers. The island's eastern shoreline, facing the main section of the river's channel, was draped with jagged layers of quartzite.

People visited the island to soak up the setting's unique charisma. Where else on the Northern Plains was there such a place? Where else could a visitor admire such a rugged, dazzling scene? Proximity made it impossible to separate the river's cascades from the island; they were part of the same geographic feature.

In Sioux Falls' youngest years, Seney Island was known as Brookings Island, named after city pioneer Wilmot Brookings, an early owner of the island. It is possible that not long after Millard and Mills moved on from the cabin they'd built on the island, Brookings and his friend Josiah Phillips shared this same shelter while more comfortable quarters for each of them were being constructed. When Brookings sold the island to George Seney in 1879, as part of the land package to build the Queen Bee mill, there was an unofficial name change to Seney Island.

That same year, a group of Minnesota and Illinois newspaper editors visited Sioux Falls. Their experience was reported by the *Minneapolis Star Tribune*. "In natural beauty and picturesqueness," reported the newspaper, "the Sioux Falls far exceed those of St. Anthony. Right close to the business part of the town is a beautiful island of about a dozen acres, covered with a magnificent growth of forest trees, which the citizens design to purchase and convert into a park." Apparently, discussion by Sioux Falls residents was underway to create a city park on the island.

The island's western shore was divided from the mainland by a narrow, ephemeral channel that during dry times and low flows lost watery contact with the main section of the river. When the western channel carried water, wooden-plank bridges permitted visitors and horse-drawn buggies to cross from the mainland.

Wilmot Brookings and, later, George Seney permitted the public to use their island property as an informal park. Religious congregations assembled there. There were political rallies and public forums. During Fourth of July festivities, there were ballgames and three-legged races. Firefighters competed against quarry workers or policemen in tug-of-war combat. Families went to Seney Island for outings. Boys courted girls. It was a dress-up kind of place.

During the city's first several decades, residents easily overlooked the fact that the island belonged to someone other than the City of Sioux Falls. They couldn't be blamed for their presumption. Seney Island, after all, was part of the city's lore. It was a public treasure. And it had been that way since the city was established. How could anyone think that Seney Island wouldn't forever be a charming, wooded isle?

Small bridge to Seney Island. *Courtesy of Siouxland Heritage Museums.*

An early debate about the future of Seney Island accompanied an 1879 proposal by the Sioux City and Pembina Railway to build tracks and a depot near the island. Hungry for additional rail service, many in Sioux Falls supported the railway's proposal. But not James Drake. The man who would become one of the leading businessmen using the falls for industrial purposes warned the city about the impacts of the new railroad development on the cascades and the island. Writing from his home in St. Paul, Drake advised the city to purchase the island "to preserve it against any encroachment calculated to mar its beauty as an attraction for visitors and a resort for your townspeople." Drake also issued a crisp bit of council. "I am bold enough," he wrote, "to predict that you will live to deplore this decision." Within several years, the rail company ceased to exist, temporarily ending the threat.

By 1887, a realization had finally sunk in: to preserve access to the island and to protect the island's character, Sioux Falls needed to create a bona fide public park there. City parks were an innovative concept at the time

During Sioux Falls' early years, families and community groups flocked to the forests and meadows of Seney Island. *Courtesy of Center for Western Studies, Augustana University.*

and had quickly gained popularity and cachet. New York's Central Park, widely viewed as a groundbreaking and quintessential urban outdoor refuge, opened in 1876. Sioux Falls aspired to join the ranks of communities on the rise. Adding a genuine park to the community's amenities would be a big step forward. Not only did residents value the island for its scenery and as a place for leisure and recreation, but a new and appealing variable also arose.

Leaders of the Methodist Church wanted to move their summertime convention for congregants from Big Stone Lake, in northeast South Dakota, to Seney Island. One factor was the island's nearness to the new Methodist college in Mitchell. Another was the location's inspiring beauty. The religious gathering would attract thousands to Sioux Falls for a week or two each summer. City leaders pondered the opportunity. Picnics, Sunday strolls and local get-togethers were pleasant, but attracting droves of out-of-towners to the island and the community would be an economic bonus. Representatives of the church and the city met with George Seney's representatives and tried to negotiate terms of a purchase but were unsuccessful. The city council felt the price of $10,000 was too steep.

In 1903, nine years after George Seney died, the new owners, a group of New Yorkers, approached the City of Sioux Falls with a proposal: find a tenant or a buyer for the Queen Bee mill, and the owners would donate Seney Island to the city for use as a park. The price tag for the Queen Bee and forty-seven acres of land contiguous to the mill structure, including Seney Island? $60,000. The city hesitated. Then discussions stopped.

That summer, the Methodists held a ten-day gathering on the island. It was a crowded event, with tents pitched everywhere. Although control of the island remained in the hands of easterners, Sioux Falls residents continued to use the setting as if it were a public park.

Also in 1903, the powerful Chicago, Milwaukee, St. Paul and Pacific Railroad, popularly known as the Milwaukee Road, began planning an expansion of its facilities west of Seney Island that would crowd the island's western channel. The natural character of the island would be compressed and compromised.

Seney Island's New York owners then proposed that Sioux Falls remove taxes levied on the island and permit the owners to charge a rental fee to groups and event sponsors using the island. In exchange for meeting these conditions, said the New Yorkers, the city could develop and maintain a park there. The city council refused the offer, and another opportunity evaporated.

Reacting to the lack of progress regarding the creation of a park at the island and the ongoing threat of industrialization, a local women's club, the Ladies History Club, pushed hard to protect the island and began to organize cleanup and beautification activities there. Their efforts, no matter how spirited and deeply felt, would be challenged and continually frustrated by more powerful forces.

The year 1907 proved to be a pivotal one for the island. Utility magnate Henry Byllesby launched his aggressive plan to harness the river and tap electricity from the falls. The straitjacket his engineers imposed on the river would block Seney Island's western channel from the main channel, altering the island's hydrological dynamics. When water stopped flowing through the western channel, the island became connected to the mainland by what was then a dry or moist dip in the landscape.

Coal-powered locomotives produced large quantities of waste, particularly cinders and clinkers left over from incinerated coal. At strategic locations, locomotives arrived above a drop pit, a long, narrow trench beneath the track, and released coal refuse from their ashpans. In Sioux Falls, some of that debris was used as ballast, but at rail sites near the western shore of the falls, some of it was scooped up and discarded onto Seney Island or into the island's forsaken western channel. Residents and businesses looking for a convenient place to dump their trash began to use the old channel. The island's forest was thinned by woodcutters. Quarry activity continued to decimate rock formations along the falls. One resident wrote about his 1912 visit to the falls and the island: "Why the city of Sioux Falls would ever have permitted this…[is] beyond our comprehension."

Vagrants and ne'er-do-wells drifted to the island. There was secrecy there, among the surviving trees and bushes. Regular citizens visiting the place were intimidated by the criminal element, and a crusade led by the police department's so-called strong-arm squad to drive off rough characters was soon underway. But nothing official was done to preserve what was left of Seney Island, and natural conditions continued to deteriorate.

The worsening condition of Seney Island became a local issue again in 1923 when the Milwaukee Road pitched a proposal to Sioux Falls. The railway company wanted to create a switchyard and expand other rail facilities near the western shoreline of the Big Sioux River, not far from the falls and bordering the diminished island. Condemnation of private property would be necessary. A bridge would carry train tracks over the river south of Sixth Street before those tracks proceeded north. Phillips Avenue ended at Fifth Street, and automobile traffic moving north or south in the neighborhood used what is now called Main Avenue. The new tracks would block plans to extend Phillips Avenue northward. Defending its proposal, the railroad suggested that Sioux Falls would never surpass Sioux City as a regional transportation hub if city government didn't allow them to pursue the expansion strategy. That warning struck a nerve, as Sioux Falls leaders viewed Sioux City as a rival threatening their community's growth and success. The city council passed a measure permitting the rail plan.

Once again, leading the opposition to a business proposal impacting the falls and Seney Island was the local Ladies History Club. "Whereas," resolved the organization, "the construction of this proposed switch yard encroaches so much upon that portion of the river generally known as 'Seney Island' and 'The Falls' as to constitute a very material damage from the viewpoint of scenic beauty and interest." Members pressured city leaders and lobbied the city council. Controversy over the railroad issue inspired what could be called the city's first bitterly contested environmental dispute.

Former Senator Richard Pettigrew also protested the proposal, complaining that the railroad's development would spoil the west side of the river near the falls and what survived of Seney Island. "If this happens," Pettigrew advised city leaders, "the time will come when the whole population will curse you for it and will probably pay some millions to get them [the railroad] out of there."

Pettigrew wrote an impassioned letter to Sioux Falls Mayor Thomas McKinnon, suggesting that no other city in the country would allow such a development to occur in such a place. He ridiculed the railroad's claims about needing the land to boost the city's economic fortunes and proposed a park be developed on the same tract. Other railroad opponents, led by the Ladies History Club, circulated petitions and collected signatures to refer the council's decision to a public vote.

Railroad supporters dismissed project opponents as "sentimentalists." *What are we fighting about?* asked the railroad company. "As a matter of fact," declared the company's attorney, "there is no Seney Island and there has been no such place for many years."

An op-ed in the *Argus Leader* scolded railroad opponents and Seney advocates: "[The island] has not been safe for a respectable woman to visit in years. High weeds, stagnant water, and a certain choice collection of tin cans and refuse detract from the scenic beauties of a place which has become an eyesore to Sioux Falls and the habitat of hoboes and prostitutes. If Seney Island can really become a political issue in Sioux Falls, then indeed the silly season has come."

The city's board of park supervisors, overseeing an agency created only eight years earlier, entered the fray. In correspondence to the city council, park officials reminded commissioners that there was interest in preserving the remaining scenic values attached to the falls and Seney Island. "Recommendations have been made from time to time for the acquirement and preservation of these scenic spots for public use," explained the park board, before urging protective action. "It has never seemed expedient to acquire and preserve these properties for public benefit until now. Through industrial exploitation, the natural beauties of Seney Island have been completely obliterated, and with every indication that the 'Falls' will meet the same fate."

The fate of Seney Island looked dismal when the citywide election held April 20, 1926, revealed the community's priorities. Of 5,549 votes cast, only 1,624 (less than 30 percent) rejected the city's supportive agreement regarding the railroad's proposal.

The railroad company completed its upgrades and eventually acquired the entire island, causing the condition of a once revered place to topple into an irreversible tailspin. Mounting industrialization imposed more and more degradation. Seney became an unsightly, dangerous landfill. Construction rubble, chemicals and old batteries from a battery recycler were randomly discarded, seeping poisons into the river and worsening toxic conditions caused by coal waste. Household trash was casually dumped. Animal parts from meat-processing facilities were strewn about. Worn-out tires were torched on-site. The odor of burning rubber replaced the scent of flowers and splashing water. Soon, the island's forest succumbed to logging. Where families once loafed in shaded grass was a mantle of reeking debris under a broiling sun. What was once Sioux Falls' favorite green space had been devastated and abandoned.

Years later, historian Charles Smith confirmed what many already knew. Seney Island was no more. Environmental problems had worsened. New problems had arisen. And no problems had been addressed. Smith complained that the island "had been profaned and ultimately destroyed by the wanton hand of avarice." His eulogy was succinct: "I consider the loss of Seney Island a crime." The City of Sioux Falls and most of its citizens, Smith grumbled, had tacitly or formally consented to this tragic destruction.

Chapter 7

LILY OF THE WEST

It happened during the summer of 1960, a year after Wayne Fanebust graduated from high school. Fanebust and a girlfriend were motoring around Sioux Falls in Fanebust's car during the wee hours, searching for a quiet place to do what young couples sometimes do in cars late at night.

"We found an empty street without houses or lights," remembered Fanebust. "I wasn't familiar with our whereabouts, but it seemed like the type of place we were looking for."

Soon, the dark gravel road they'd turned onto led them into a tangle of trees, a secluded setting that was equally spooky and appealing. This street, Fanebust thought, was like no other street he'd seen in Sioux Falls.

To gather his bearings, Fanebust stepped out of the car. In the distance was the glow of the local meatpacking plant. Fanebust had a general idea of where he was. He could hear the loud hum of moving water, perhaps the Big Sioux River. He hiked toward the sound and soon saw a waterfall. A waterfall!

Spread out before him and visible in the moonlight was shimmering, turbulent water dropping over rocky ledges. *What is this place?* Fanebust wondered. There was lots of stone arrayed like a bumpy patio or in blocky stacks. The setting was unexpected—and wildly inconsistent with the surrounding cityscape.

Surprisingly, Wayne Fanebust hadn't heard of the falls before he looked upon them. Growing up in and near Sioux Falls in the 1950s, Fanebust had not been shown the cascades of the Big Sioux River by anyone: not a teacher, a parent or a trouble-seeking posse of his pals. Even ordinary boys seem to find the most obscure and risky places in any town. But there stood Wayne Fanebust, a young man who claimed Sioux Falls as his birthplace, and he'd never before encountered the community's namesake.

Fanebust later became an attorney and wrote books about Sioux Falls and people who shaped the city. "It reveals a lot about what happened to the Big Sioux River and the falls," said Fanebust, "that so few knew much about the place and that so few visited the falls."

Most of the leading developers who established industrial businesses along the shoreline of the falls were long gone. No matter their human impermanence, the evidence of their ambitions remained. Environmental impacts typically outlast the men who cause them.

George Seney died from heart disease in 1893 at age sixty-six while resting in his sumptuous quarters at the Grand Hotel in New York City. Remembered as a financier, developer of railroads, collector of fine art and generous philanthropist, Seney left behind the Queen Bee mill, an industrial monument in a remote town that reflected the adventurous drive of a risk-taking member of the monied class. The Queen Bee stood high and mighty at the time of Seney's death, and it would serve a variety of increasingly unimportant purposes during much of the rest of its life.

At one time, James Drake employed hundreds of men at his Sioux Falls quarry and rock-polishing factory. His buffing business had been a phenomenal success, the most nationally notable enterprise using the falls. Drake's entrepreneurial pursuits advanced the city's economic fortunes during critical early years of settlement. Developing rail service and finding potable water for Sioux Falls were his other priorities. Drake also embraced the importance of creating a park at Seney Island and advised the city to do so. His counsel was ignored, setting the stage for a sequence of rail and other commercial developments near and over the falls and Seney Island. Drake died in Pasadena, California, from an apparent heart attack in 1912, at the age of sixty-seven.

Henry Byllesby brought accomplished scientific and business savvy to the falls. His electric utility on the Big Sioux shoreline marred the scenery, wrecked a long stretch of channel and polluted the environment. He viewed the natural falls as expendable, a source of power to advance society by providing electricity. Byllesby's impact on countless communities across the United States and other nations was profound. At one time, he and his team oversaw more than one hundred facilities generating electricity, most of them in the Upper Midwest. Byllesby contributed his organizational skills while serving overseas during World War I. Approaching age sixty and independently wealthy, he volunteered to serve as a purchasing agent for U.S. military operations based in London. Following the war, he was awarded the American Distinguished Service Medal by the United States, and England later bestowed the Distinguished Service Order on him. A heart attack killed Byllesby in 1924 at age sixty-five, during a visit to his dentist.

Sioux Falls legend Richard Pettigrew was known to his closest friends as Frank. The group of confidants who used that moniker was probably small but included some of the most renowned and influential people of his time. Intense and combative, Senator Pettigrew seemed to alienate people as often as he attracted them. Viewed as a guy willing to take on the status quo, Pettigrew was surprisingly progressive, supporting labor unions, women's suffrage and municipally owned utilities. He pursued business deals with single-minded fervor and claimed the principal objective of his work was to create jobs and grow Sioux Falls. He believed that service to industry was the chief purpose and benefit of the falls. Despite his free enterprise exploits, Pettigrew became a socialist in the final years of his life. In addition to his significant political and public service, he was an accomplished author and unyielding social critic. He died in 1926 at age seventy-eight from a probable stroke, and his remains are contained in a private mausoleum he built for himself and his family in Woodlawn Cemetery, a seventy-acre parcel he donated to the City of Sioux Falls in 1903 for use as a cemetery.

What did these accomplished men see when they gazed at the cascades of the Big Sioux River? Byllesby, Drake, Seney and Pettigrew contributed in countless ways to a world they sought to master. They were exemplary citizens who lacked environmental awareness and sensitivities. Their vision for the cascades highlighted economic opportunities, and Sioux Falls agreed with their objective and encouraged it. Most in the community had come to view what happened to the cascades and Seney Island as acceptable collateral damage. Commerce, it was believed, was more consequential than sustaining nature.

The emergence of women as community activists had long been bubbling through the muddy resistance of conventionality. The impact of this important movement was boosted by the advent of women's clubs across the nation as early as the 1860s. Those early women's clubs emphasized fellowship and self-improvement.

Sioux Falls' first women's group formed in 1879 for the purpose of studying literature and history. Its founder was thirty-five-year-old Reverend Eliza Tupper Wilkes, a college-educated Unitarian minister who helped establish Colorado College, a coeducational liberal arts institution, and many churches across the Upper Midwest, including Sioux Falls. The loose-knit handful of women Wilkes convened in Sioux Falls called themselves the Ladies History Club. After casually meeting for a couple of years, the club decided to formally organize and choose officers. Membership had reached thirty by 1885, and Mrs. Richard Pettigrew was elected president. Reverend Wilkes became prominent in the women's movement, serving as a vice president of the National Woman Suffrage Association in 1884 and sharing the speaker's dais several times with Susan B. Anthony.

At the turn of the nineteenth century, the history club endured internal tension over its purpose as an organization. One faction within the group was eager to engage in civic affairs and public service. They'd publicly protested the degradation of Seney Island and supported creation of a park there. Another faction preferred the club offer only social and educational opportunities for its members. Eliza Wilkes had by then moved to California, though it seems certain she would have been pleased that the activists prevailed.

Mrs. P.H. Edmison, the club's president from 1914 to 1919, was proud of the new focus. "There is no such thing as standing still," she declared. "When growth ceases, decay sets in."

Under Edmison's leadership, the organization changed its name, becoming simply the History Club. And membership doubled, confirming the wisdom of the decision to reorient the group. The club came to comprise different departments, each focused on specific themes, such as public health, conservation and schools.

By the 1920s, the General Federation of Women's Clubs (GFWC) was the United States' largest women's organization, and the group's goal was straightforward: to shape and improve communities, states and the nation. Hundreds of different women's clubs from communities in every corner of the country joined GFWC to share information, tactics and support. Among them was the History Club in Sioux Falls, where membership had swollen to at least several hundred women, making it the largest women's group in South Dakota.

Conservation and outdoor topics were of special interest to women's clubs. Mary Belle King Sherman served as chair of GFWC's Conservation Committee from 1914 to 1920, and her vigorous support for creating and protecting national parks earned her the nickname National Park Lady. The organization also fought a long, bitter and losing battle over the proliferation of billboards along America's roadways.

Likewise, one of the departments within the History Club focused on conservation issues, and programs were presented about the value of wetlands and pollution threats from agricultural insecticides. At the time, men's groups seemed disinterested in such topics. The club's 1923 campaign to block railroad development near the falls and Seney Island had been unsuccessful, but the women gained experience and visibility. The organization's membership exploded, reaching three hundred by 1928. The club sought to raise awareness about the falls and organized trash collections and tree plantings there. The organization became an incubator of public-spirited activism.

Identifying a date signifying an official start to Falls Park is as confounding as Sioux Falls' inconsistent relationship to the falls. There was, it seemed, no groundbreaking event, no ribbon cutting, no gathering of city leaders clutching spades and making speeches. But a sequence of important events inched the city forward.

A Watertown, South Dakota quarry company owned property immediately north and west of the falls, and in 1932, the company asked the City of Sioux Falls to relieve its tax burden on that land. After some unproductive give and take, the city purchased the property. This was heartening news to those dreaming of a genuine park at the falls. The land needed to pursue that dream was now owned by the city.

The following year, park supporters were encouraged when the city developed a road leading sightseers to the falls. No longer, said one observer, were visitors forced to "tramp across rocks, through cactus strips, and over rugged territory." From the comfort of their car, a family could drive over a gravel roadway to view the river and its cascades up close. Watching a patch of nature through a windshield was, apparently, a sufficiently meaningful outdoor experience for many.

N.E. Hansen, professor of horticulture at Dakota Agricultural College (which later became South Dakota State University), visited the new informal park bordering the western side of the falls in 1933. His enthusiasm was accompanied by advice that future park improvements should emphasize nature and native plants. All plants not native to the area should be removed, he suggested. Hansen also had ideas about the future of the park. "It would be a mistake," he warned, "to commercialize or modernize the area in any way."

No doubt Hansen was troubled by what he saw around the park. Along or near the street leading to the falls were a city-owned quartzite quarry, an asphalt plant, maintenance shops for the street department, storage buildings and dusty parking lots for city vehicles. There had been no public opposition to these developments.

The city's recently acquired land joined other blemishes on the western side of the falls: shabby warehouses, garbage dumps, a junkyard, a couple factories and a railroad roundhouse and tracks. There was no evidence demonstrating a serious interest in beautification near the falls. Professor Hansen's recommendations were ignored.

The city-owned quarry near the falls combined unusual partnerships. For part of the quarry's existence, a portion of the labor came from inmates serving time in the local jail. Facing chain gang toil as a consequence of judicial conviction discouraged criminals from preying on Sioux Falls, said the local sheriff. The quarry was also part of the federal Works Progress Administration program that offered invaluable employment when job opportunities across the nation dwindled.

By 1939, the quarry employed up to one hundred men. During an especially busy year, the quarry yielded more than eighteen thousand cubic yards of quartzite. Two mechanical rock crushers were necessary to keep up with demand. A dynamite expert worked continuously at the quarry for eight years. Visiting the falls wasn't a peaceful experience.

A decade later, the so-called park at the falls, casually called Falls Park, still lacked official park designation and appropriate management. Grassroots community groups, including the History Club, the county historical society and the local chapter of the Izaak Walton League, understood that a formidable obstacle blocking their vision of a genuine park at the falls was the contradictory purposes pursued by different city officials and agencies regarding the use of the city's property along the falls. Those varying objectives, they realized, left the fate of the setting in a tenuous position. "Since nobody had the responsibility of maintaining the [falls] area," said city commissioner Bert Yeager, "it fell by the wayside."

To rectify this shortcoming, park supporters posed a question to city leaders: Which specific branch of Sioux Falls government would be best suited to managing the site? They urged that oversight be consolidated to the park board.

In May 1953, the city council voted to transfer management of the tract west of the falls to the Parks Department. At the park board's June 5, 1953 meeting, park officials agreed to accept responsibility for the property. Some might say that this action represented the official birth of Falls Park.

Acting quickly, the park board within four months released a planning document and renderings of a proposed park design.

When the Sioux Falls Parks Board assumed control of Falls Park in 1953, industry dominated the cascades vicinity. *Courtesy of Siouxland Heritage Museums.*

Acclaimed Siouxland writer Frederick Manfred was fond of the falls, and he applauded the city for investing in the park. On November 11, 1953, a letter written by Manfred appeared in the Sioux Falls *Argus Leader*.

> *For many years I've lamented the fact that the city of Sioux Falls had never taken advantage of a real scenic spot located right in the heart of town. Despite mounds of garbage around it, and debris from manufactuaries and rearing walls of industry, I always liked taking a walk past the old roaring tumbling falls. The rugged redrocks were beautiful in the sun and the falling driving water stirred me to the depths of my being. How I longed to see the falls in its former natural glory. And now I see that the city is going to make a park of it....Wonderful!*
>
> *I have another suggestion....The city should try to buy up the land a good 100 feet to either side of the Sioux River both above and below the falls, and make both banks grassy swards and meadows again, with slow, curving walks and stone benches and other picnic sites...and maybe even troubled business executives will take it into their hearts to stroll along it to walk out their troubles. Artists, poets, novelists, let alone lovers, will meanwhile hail it as making Sioux Falls, as cities go, the Lily of the West.*

A falls visitor who arrived on the river's western shoreline looked across the cascades to a hodgepodge of what Frederick Manfred called "manufactuaries." What they gazed upon wasn't beautiful or graceful like a lily, although the aspect of the falls that survived damming and dynamite retained a surprising share of its grandeur.

Conspicuous after many decades was the Queen Bee mill, unmistakable and notable by its colorful, unfortunate history. Failure had followed a cloudburst of fanfare. The building had been built mighty and had done little.

After Richard Pettigrew and his group of Queen Bee speculators fell on their faces, the mill sat idle for more than twenty years. Then came an effort to resuscitate the enterprise in 1911, when Henry Byllesby sold the facility to United Flour Milling of Minneapolis. The new owner invested heavily in the mill, modernizing equipment and adding electricity to power the operation, a pragmatic decision considering the proximity of Byllesby's utility. The upgraded mill boasted the largest electric motor in South Dakota, a five-hundred-horsepower titan. United's top-selling product, marketed as Queen Bee Flour, was shipped to East Coast and international customers for nearly six years before bankruptcy forced United Flour to transfer all contracts to its Minneapolis milling plant. Not long afterward, another Minneapolis company, Commander Larabee Elevators, Milling and Storage, resumed operations at the Queen Bee facility. The company ran the renovated mill intermittently until 1920, when they sold it to a group of Sioux Falls investors who diversified the mill's output to include oatmeal. But the cereal company sputtered, and Larabee reacquired the Queen Bee property. The one-time marvel never milled again.

In 1929, Larabee revamped the mill's vast insides to serve as a warehouse, and in 1937 the building was sold to Ben Margulies, a Sioux Falls real estate developer and car dealer who rented the former mill as storage space to local companies.

On a frigid January night in 1956, the old mill entered the next phase of its unfulfilled life. No one pinpointed the cause, but the fire that gutted the entire structure was visible for miles around and drew hundreds of spectators. An airline pilot flying overhead reported that when the structure's roof collapsed, explosive flames triggered a column of heat that rose half a mile into the sky. It should not have been unexpected when the Queen Bee's owner delivered another tidbit of odd news: the once venerated building was uninsured.

As the Queen Bee's seven-story skeleton smoldered, park work slowed, paused and then came to a complete standstill. One Sioux Falls resident

The overhyped and underused Queen Bee mill was gutted by fire in 1956. *Courtesy of Siouxland Heritage Museums.*

described his visit to the neglected park: "When we did find how to get to [the falls], the road was almost impassable, but we were amazed to see the beautiful sight that was hidden away in such run-down surroundings."

Six years after his impassioned letter to the editor supporting efforts to create a park at the falls, Frederick Manfred was upset. Very little had been done to build the park, and he implored action. This time, he did so doing what he did best. *Conquering Horse* was Manfred's eleventh book, and he declared that a central reason he wrote the novel was to inspire public appreciation for the cascades. "The falls have haunted me since the old days when I lived in Sioux Falls in 1936–1937," explained Manfred. "It must have been beautiful before the white man put up his mills and so on around it....I hope my book becomes a raging bestseller if only to stir the local people into doing something about a famous and beautiful landmark of theirs."

Various sections of the book take place at Sioux Indian campsites near the falls. One memorable scene describes teenage Sioux boys climbing a rocky cliff at the falls to use a bathing area reserved for them. They cleanse themselves using fine quartzite sand that collected in a calm, stone-sided pool. Above the main cataract, Manfred writes, the river "spilled across staggered slabs of rock in a thousand tiny streams. Eons of flowing water, grit-laden, had honed the red quartzite down to so fine a polish that it resembled the smooth silken flesh of quartered beef." The protagonist in *Conquering Horse* refers to the Big Sioux River as the "river of the double bend" and to the cascades as "falling water."

Conquering Horse, published in 1959, did become a popular book. Rave reviews were plentiful, including a *New York Times* appraisal declaring it one of the year's finest novels. Celebrated filmmaker Michael Cimino, who later won an Oscar for directing and cowriting the cinematic masterpiece *The Deer Hunter*, attempted to adapt Manfred's tale for the screen, but budget issues halted the project.

Manfred wasn't the only artist or activist inspired by the falls. Public sentiment for the park was reclaiming momentum. A courageous leader was needed.

Chapter 8
HAZEL

Barbara O'Connor recalled the first time she encountered her future mother-in-law. Barbara was dating Michael O'Connor, and her boyfriend asked her to meet his mother. Barbara arrived at the O'Connor home in central Sioux Falls and was directed to the kitchen. Standing in front of the fridge, with one foot raised directly above a metal can placed upright on the linoleum floor, was Mike's mom, Hazel. As Barbara watched, Hazel stomped the can, crushing it with one blow. Then she stooped down, snatched the flattened container and tossed it into a trash bin.

Hazel noticed that her young visitor looked puzzled. "She explained to me that the Sioux Falls city dump was running out of room," Barbara remembered. "She said everyone must reduce the volume of trash they send to the landfill." Barbara had never pondered the condition of the local landfill before.

In the late 1940s and early '50s, few worried about overflowing garbage dumps, but Hazel O'Connor did. The stewardship philosophy of a raw and novel environmental movement had not penetrated mainstream American society and wouldn't for many decades. The concept of conservation was in its infancy. But O'Connor and other like-minded citizens in Sioux Falls, especially women who were members of the History Club, were the portents of times to come.

Hazel O'Connor was born a Lundquist in 1897. Her mother, Theresa Lundquist, left Sweden for the United States alone at age nineteen, eventually earning her keep as a seamstress in Boston and learning English from American magazines. Theresa read about a Swedish stonecutter in Illinois named Elof Lundquist and wrote to him, asking if they were kin. Elof responded, denying any relation, but a four-year correspondence began. When Elof moved to Jasper, Minnesota, to quarry Sioux Quartzite, he asked Theresa to join him, and in 1890, the couple married and settled in Garretson, South Dakota. Five years later, the couple moved to Sioux Falls, where craftsmen were needed to build a large federal courthouse constructed of Sioux Quartzite. Elof Lundquist's skills secured for him a permanent occupation, and soon the Lundquists settled into a comfortable home in east Sioux Falls, near the quarry where Elof was employed. Hazel was born the following year.

Hazel admired her father's handiwork that was visible on prominent buildings throughout Sioux Falls. The quartzite blocks he cut and shaped and the structures they formed and decorated reminded her of him, like warm-blooded mementoes. She remembered many visits as a girl to the cascades of the Big Sioux River. "My family always went to the falls with a picnic lunch on special holidays—the Fourth of July, circus days, Sundays," she recalled. "We went there in a surrey with fringe on top. When I was a child, I thought the falls was the most beautiful place I had ever seen."

As a sixth grader, Hazel wrote a story describing the value of public parks: "Parks furnish recreation places for the people who have no yards or grounds at their homes." Her altruism was already evident. For the rest of her life, she would admire and advocate for parks.

O'Connor had been a member of women's organizations since the 1920s, and she witnessed the contentious debates over the fate of Seney Island and noticed the deterioration of scenery and nature surrounding the falls. She was actively engaged in community affairs by the time she explained the state of landfills to her future daughter-in-law.

O'Connor and other History Club members aggressively fundraised before building and opening their own clubhouse in 1940. The property where the facility was built, 758 South Phillips Avenue, was a compact lot located near the southern edge of Lyon Park. The casual viewer might have confused the handsome, one-story brick building with a normal house in a pleasant neighborhood, and the similarities were intentional. The structure's homelike interior featured a spacious, comfortably furnished parlor beside the entryway, and a wide hallway passed a kitchen and led to a gala-sized room fronted by a sizeable stage. Along the long north wall of this handsome space was a large fireplace. Generous windows brightened the building with natural light. In this collegial setting, the club held its meetings and rallied its members. This was where Hazel O'Connor met with friends and allies to plan strategies and events to advance Falls Park.

O'Connor had surely noticed an article appearing in a 1956 edition of the *Argus Leader* written by an Iowa woman whose family regularly visited Sioux Falls. On one trip to the city, the Iowan inquired about the community's name. "Why is the city named Sioux Falls?" she asked clerks and waitresses. Someone eventually answered. When the woman and her family finally found their way to Falls Park, they were astonished. "The surrounding grounds are a bit unkempt, we must admit," she wrote. "but the thrill of finding this rose-colored, rainbow-set jewel in however rough a setting outweighed any disadvantages.…It's Super Colossal!"

Then there was the letter in a 1959 edition of the *Argus Leader* providing a stark reminder about the dismal condition of the park. The park "is a spot on the face of the city that stands out like a boil and hurts like one!" wrote the local resident. "The park through all these years? Phooey. No park at all. I have wondered when someone would wake up and see the possibilities. All this time the park has been a dump. It is a shame."

Hazel O'Connor was wide awake by then, having already implored the History Club to adopt beautification of the falls and the river as special projects. "It aggravated and hurt me," she later explained. "The place our city was named for was an embarrassment to the city."

O'Connor recognized the value of organizing a coalition. "In union there is strength," she declared to her fellow club members. And she went about expanding her grassroots troop, convening leaders and members from at least thirty women's groups to join the effort. "I hope that all of you feel that you are part of an organization that is going to work for the improvement of your community," she announced at a large gathering. "Don't get discouraged when there are reverses, but continue to work, for a thing of beauty is a joy forever."

In 1959, while serving as president of the city's Federation of Women's Clubs, O'Connor led a group of history enthusiasts to protest the planned destruction of the county courthouse and offices, a beautiful structure made of Sioux Quartzite and completed in 1890. Minnehaha County needed a new building, county officials said, because the old building's rooms were cramped and inefficient. O'Connor and her confederates forcefully lobbied to preserve the building, considered an architectural masterpiece. It was a long fight they eventually won, and some years later, the historic building became the county's history museum, a venerated institution and landmark. That triumph elevated O'Connor's reputation as a progressive, effective advocate.

As the 1960s dawned, O'Connor's activism increased. This was good news for park advocates, but for members of the Sioux Falls park board, their volunteer service would suddenly become more challenging, more scrutinized. Hosting persistent and exacting visitors at staid park board meetings had been a rare occurrence until Hazel O'Connor stood before the board and introduced herself for the first time.

Many in Sioux Falls were already familiar with Hazel. Her volunteer work for a variety of organizations, including the History Club, the South Dakota Federation of Women's Clubs, the Sioux Falls Chamber of Commerce, the Red Cross, the Izaak Walton League and the PTA exposed her to an array of citizens. She acted in local theatrical productions and presented humorous readings and stories, sometimes speaking in Swedish to add comedic emphasis. Hazel and her husband, Joe, owned and ran a successful printing business in Sioux Falls, and Joe also served civic-minded organizations. The couple raised three children.

On weekends, Hazel, Joe and Joe's two brothers and their wives played poker. Hazel was an enthusiastic gardener, a birdwatcher and a devotee of pasqueflowers. "She could grow anything," said her daughter-in-law Barb O'Connor, "and she was the heartbeat of a very close family." Hazel O'Connor was a character with character.

Larry Weires was a staffer at the Sioux Falls Parks and Recreation Department dealing with matters related to Falls Park when he first encountered Hazel. "She wasn't more than five feet tall," remembered Weires, "but when you were with her, you felt like you were in the company of a giant. She was a dynamo."

Minutes from park board meetings note frequent appearances by O'Connor. During one meeting, she asked the city to add benches, lighting and fencing to Falls Park. The park's road needed paving, she complained. The board told her that funds were limited and spread to parks throughout the city. But she was undeterred. "Why aren't more funds committed to this meaningful purpose?" asked O'Connor. She issued more complaints, more requests, more suggestions. "What about walkways for visitors?" she queried. "How come there aren't more plantings? Why isn't the city removing quarry rubble and other trash from the park? Why isn't there a caretaker? What about helpful signage?"

During another meeting, park board members explained they were unable to pursue certain improvements at the park. O'Connor's impatience was noted. "Mrs. O'Connor was most vehement in her criticism of the board for what she regards as a lack of interest and action," wrote the board's

Community activist and Falls Park champion Hazel O'Connor (*right*). *Courtesy of Siouxland Heritage Museums.*

secretary. Board meeting minutes were typically bland and dispassionate. Hell-raising Hazel hassled the status quo.

O'Connor often appeared at park board meetings accompanied by collaborators and friends, but she was almost always the featured park champion. Her coalition embarked on a park and river visibility campaign, sponsoring boat races, public programs and park cleanups. There was hardly a group or club in the city that didn't hear O'Connor's pitch about improving the park. Slowly, the community was being reintroduced to a misused resource. Slowly, the City of Sioux Falls was responding to pressure from its citizens.

Earl McCart had served two terms as Minnehaha County auditor when he decided to seek a five-year term as a Sioux Falls city commissioner in 1960. At the time, there were only two city commissioners and the elected mayor at the top of city government, so commissioners wielded considerable influence and power. McCart might have been the first candidate for Sioux Falls city government office to highlight Falls Park as an issue worth tackling, and he didn't mince words when describing the park. "If the entire city of Sioux Falls was in the same condition as the site after which this city is named—the area around Falls Park—we could well be termed the 'junk heap' of the Upper Midwest," he declared in one campaign advertisement. McCart's proposals were reasonable: Clean up unsightly debris from quarries and from the park in general. Fence off and obscure nearby city garages and storage buildings. Plant a screen of trees to conceal junkyards in the park area. Add picnic tables to the park. These projects mirrored much of what Hazel O'Connor was suggesting.

McCart's narrow victory would prove consequential for Falls Park proponents. The Sioux Falls native was in the first months of his term when Hazel O'Connor demanded that every dollar of the $5,000 budgeted by the city for the park during 1961 be spent. O'Connor insisted that not spending available money was unacceptable. "We intend to keep the pressure on," O'Connor vowed, "and we are going to ask for more money for 1962." Superintendent of parks John Browning told O'Connor that his department was receiving pushback from citizens more interested in golf courses, horseshoes and tennis courts than developing the park at the falls. As a city commissioner, McCart functioned as the commission's liaison to the park board and also served on that board. He listened as O'Connor told parks and recreation directors that investing money in golf courses was less important than investing at Falls Park.

O'Connor's badgering paid off. At a meeting of Sioux Falls women's clubs, Superintendent Browning announced the city would add concrete viewing areas near the falls and blacktop the park's roadway. Better parking and nighttime lighting, said Browning, were also in the works. The viewing areas Browning promised were completed, and they remain functional after many years, presenting themselves as iron-railed sidewalks built atop the rocky western shoreline of the falls.

Browning also suggested that park enthusiasts must consider the far-off future of the place. "What sort of character should our park present?" he asked. The city, he said, would one day need to determine if it wanted a park emphasizing "rustic beauty" or a heavily landscaped park "in a more modern trend." Browning's appeal to that audience was prophetic. Rustic versus modern would long be a debated topic.

One meeting attendee, Mrs. Joseph Nelson, representing the East Side Community Club, suggested that the city purchase property on the eastern shore along the falls. Public access to the falls, she reminded everyone, was restricted to the west side of the river. Nelson explained that once eastside property was acquired, it could be cleared out and cleaned up, and not only would the falls be accessible from both sides of the river, but a more picturesque view from the west side of the falls would also improve the visitor experience. This may have been the first public overture promoting park expansion on the east side of the falls.

The year 1962 brought unexpected tragedy to Hazel O'Connor and her family. While she and her husband were returning to Sioux Falls from a road trip to see their children in Arizona and California, Joe was admitted to a hospital in Wichita, Kansas, where he died. Michael Joseph "Joe" O'Connor was seventy-four years old. The printing business he founded in 1936 carried his family name, and after his death, the company's management was assumed by his only son, Michael, who later served in South Dakota's legislature and ran unsuccessfully as a Democrat for governor.

Hazel's close friend Carol Mashek said O'Connor's activism intensified after her husband's death. "She turned the energy of bereavement into a positive force," remembered Mashek. The world was awakening to the importance of a healthy environment, and that pleased O'Connor, but she understood that defying conventionality was a heavy lift, especially as she confronted business practices in her hometown.

Sioux Falls Corrugating Company was established in 1918, operating in a sixty-foot-by-sixty-foot building situated not far north of today's Sixth Street bridge, along the west side of the Big Sioux River and a short distance upriver from a stretch of the cascades known as the upper falls. Three men from Wisconsin, including Charles Rysdon, had selected Sioux Falls for their new business. They chose wisely and managed skillfully. Soon, their sheet metal fabricating company outgrew the original building. By 1925, the company was manufacturing and selling ventilating pipes, furnace fittings, metal roofing and siding, oil tanks, watering tanks and other agricultural supplies to customers in seven states. A series of building expansions was necessary to keep up with demand, and the company's property along the river grew increasingly congested.

In 1937, the company changed its name to Sioux Steel, and additional growth reflected savvy ownership and swelling markets. Patents were developed and the product line expanded. By 1955, the company employed two hundred people and shipped metal buildings and grain bins in pieces for assembly on location. In only six minutes, Sioux Steel's automated factory could produce the raw components necessary to construct one grain bin. But Sioux Steel's commercial successes meant its need for additional facilities and space became more acute. Property lines confined the company on three sides. On the fourth side, the eastern side, was the Big Sioux River.

Big Sioux River shoreline at Sioux Steel, circa 1950s. Hazel O'Connor and other river advocates clashed with the company over aesthetic and environmental issues. *Courtesy of Sioux Steel.*

Where Sioux Steel's land abutted the channel, the river was wider than usual. The company decided to convert a portion of this stretch of river into usable terra firma by dumping fill directly into the channel. The expansion project wasn't trivial. Reports described the company's conversion of river into land as measuring up to ninety feet from the original, natural shoreline, creating two new acres of property and smothering a large section of a public waterway, including at least a portion of the western channel that had once skirted Seney Island.

Hazel O'Connor wasn't pleased, vigorously opposing what she described as an "encroachment" into the river. She was joined in her opposition by other Sioux Falls citizens, particularly members of local women's groups.

At one testy meeting, Sioux Steel president Max Rysdon, son of company founder Charles Rysdon, told Sioux Falls officials that he was surprised and disappointed the community might contest his company's actions. "Do you want us to expand in Sioux Falls or not?" he asked. Rysdon explained that expansion would mean 250 more jobs for the city. He noted that other communities were courting the company, offering up to ten acres with free water and sewage systems. Sioux Steel claimed its incursion was actually good for the river, providing a necessary "straightening out" of the channel.

Rivers were mostly unregulated, although the Rivers and Harbors Appropriations Act of 1899 sought to prevent the deposition of materials that hampered commercial navigation. The Big Sioux, of course, was not a navigable river. Forty years would pass after Sioux Steel's brazen expansion tactics before regulations were enacted to explicitly prohibit what the company had done. Permits would be required, and decision-making, monitoring and enforcement became the responsibility of the U.S. Army Corps of Engineers, viewed as the nation's river management experts.

But in 1963, the City of Sioux Falls admitted its inability to deal with Sioux Steel. Max Rysdon proposed his company assume ownership of the newly created property. Sioux Falls city attorney John McQuillen explained that title to the bed of the river belonged to the State of South Dakota. "We can't give you property owned by the people," said McQuillen. Commissioner Earl McCart told Rysdon that the city would do anything to help the company "except give you the river." Mayor V.L. Crusinberry supported a hydrologic examination to determine how Sioux Steel's incursion impacted river flows.

The U.S. Army Corps of Engineers was asked to evaluate the situation. Although it concluded that the company's actions had a negative impact on the character and behavior of the Big Sioux, the agency claimed it lacked the budget and authority to remedy the issue.

Hazel O'Connor sidestepped city officials; phoned South Dakota's attorney general, Frank Farrar; and asked that Farrar's office force Sioux Steel to remove the fill. She had already circulated a public petition requesting this resolution and presented that petition to Sioux Falls city commissioners.

Farrar explained to O'Connor that his team had been investigating the situation. "[We]…will determine the rights of the public in the river and the waters flowing therein in addition to determining what rights the adjacent owner has in the use of the river," stated Farrar. "It would, of course, be the state's position that the Sioux Steel Company is encroaching upon public lands and waters and that such encroachment must be abated."

By late August 1964, the matter remained unresolved, and the attorney general's office told O'Connor that the investigation was ongoing. In September, Assistant Attorney General L.A. Weisensee wrote to O'Connor, informing her that a meeting with Sioux Steel had been held. "This matter involves serious and complex questions," said Weisensee, before inviting O'Connor to phone the attorney general's office to learn more and share her views. O'Connor the activist must have been deeply frustrated by the slow pace and the government's inability to oversee and settle the matter.

Attorney General Farrar later revealed that his office considered two options to resolve the problem. His office could force Sioux Steel to remove the fill, or the land could be transformed into a city park. The company agreed to give the artificially created property to the State of South Dakota, and the state then deeded the two-acre parcel to the City of Sioux Falls. Local Kiwanis Club members planted grass and trees on the property, and the small plot became known as Kiwanis Park. The park had confusing access through the Sioux Steel complex and sat alongside a cluster of industrial buildings. Its popularity and recreational usefulness never materialized.

Better news came from the east side of the river in December 1963, when Sioux Falls real estate developer Ben Margulies donated the Queen Bee property to the city. Commissioner McCart sought input from Hazel O'Connor regarding the mill. Some city leaders supported demolition and removal of any trace of the structure. O'Connor and McCart agreed that the old mill could provide historical education to park visitors. "Clean up the rubble," McCart told other city leaders, "but leave the walls standing." McCart's second suggestion was to place a viewing platform on top of the stone foundation of the former mill's turbine house. Both ideas were adopted by the city.

Quarry pits on city property adjacent to and within the boundaries of Falls Park remained unsightly and unsafe. To conceal and cover these excavations, local garbage haulers and the general public were encouraged to discard trash and waste in the gaping, rocky cavities.

Richard Burns's family owned a Sioux Falls trucking company, and as a teenager in the early to mid-1960s, he was tasked with cleaning out warehouses on weekends. The rubbish he collected was delivered to quarry pits at the park. "There were two or three deep quarry pits west of the river in the area where Falls Park sits," recalled Burns. "Sometimes I'd back right up to a pit and throw junk straight into a hole." When an abandoned quarry was stuffed with garbage, the city added a layer of black dirt so grass could be grown, erasing any sign of past disturbance. This coarse approach to beautifying parkland marked another improvement to the park, no matter how callous it was.

It became apparent to O'Connor and others that a community-based organization focused on the falls and the river must be formed. This inspired the Sioux Falls River Improvement Study and Evaluation Committee, created in 1966, which later shortened its name to the River Improvement Society. Starting with an ungainly name, the group eventually adopted a catchy acronym, RISE. Joining O'Connor as incorporators of the new nonprofit were John Gerken, a local banking executive, and hardware store owner Roy Nyberg.

The year 1969 marked the debut of an ambitious plan for the Sioux Falls waterfront, a concept proposed by consultants from the University of

Nebraska School of Architecture and supported by the City of Sioux Falls and RISE. The consultants' report noted the blighted condition of the Big Sioux shoreline and its "negative influence" on the city's commercial district. The Nebraskans were optimistic that the riverfront could one day serve a "vital role in the future growth and development of the City."

Concepts in the plan included a soaring edifice, called the Sioux Tower, to mark the entrance to a series of parks along the river. Those parks would have specific themes: an equestrian park, a science park and an "Oriental park" to celebrate Asian culture. There would be an open-air theater, a water sports center and a land sports center. Visitors could overnight in a campground. "These elements," wrote the planners, "will be integrated into a system of activities and connected park drives, pedestrian ways, and bicycle paths." There was, RISE acknowledged, much to consider. The proposals were a bit overwhelming.

But Roy Nyberg was unfazed. He believed a sophisticated river project in San Antonio, Texas, was a venture worth emulating, and he helped sponsor a visit by community leaders to assess that city's fabled River Walk. Often cited as a model for urban development, the River Walk bordered a stretch of former river channel bypassed by a flood control project in the 1920s. The city wanted to replace the abandoned channel with a sewer system, but local preservationists protested, and a different concept emerged. The discontinued channel became a concrete canal linking many of the city's historic sites and buildings. Flows were carefully controlled and kept at an agreeable level. Wide sidewalks and decorative bridges flanking and crossing the two-and-a-half-mile waterway were built. Lush landscaping was installed. Stunning architectural development followed. A sparkling entertainment district prospered. The unique waterfront neighborhood became the pride of San Antonio and drew visitors from across the country.

By 1972, Sioux Steel and its relationship to the Big Sioux River were back in the news. This time the squabble between citizens and the company revolved around the company's use of its own property near Falls Park. To expand a gravel parking lot and loading areas, Sioux Steel cleared brush and trees. Hazel O'Connor was incensed. She accused Sioux Steel of creating an eyesore near a city park and on the former Seney Island site. City commissioner McCart criticized the company for dumping trash on land adjacent to the park. One member of the park board, Mrs. Duane Smith, said she was "sickened" by the scene.

Hazel O'Connor wrote to South Dakota Attorney General Kermit Sande, requesting that his office block Sioux Steel's plans. Sande advised

O'Connor to work with the City of Sioux Falls on the matter. Earl McCart and the city's parks and recreation board passed a resolution instructing the city commission to acquire Sioux Steel's modified property through condemnation.

Sioux Steel told city commissioners that condemnation of the cleared land would be a "death blow" to its business. The company's attorney asserted that while many South Dakota cities encouraged industry, this type of interference was unusual. "We hear more about [industrial development] than ecology," he said. Was it reasonable, asked Sioux Steel supporters, that the city could dictate what the company did to its own land? In this instance, said Hazel O'Connor, the answer was yes. The successful manufacturing company, she contended, was proving to be a lousy neighbor to a city treasure.

Despite McCart's opposition and Hazel O'Connor's tenacious criticism, the city commission voted two to one to reject the condemnation resolution. Sioux Steel continued to adapt its limited property to accommodate escalating customer demands and also purchased new space and facilities elsewhere, including a business and factory in Hull, Iowa, and another plant in Lennox, South Dakota.

Falls Park proponents celebrated several new land donations on the eastern side of the river. Northern States Power (NSP) had produced electricity from alongside the falls since 1916, but the utility had opened a larger generating facility in northeast Sioux Falls in 1949 and ceased hydropower production at the falls in 1961, citing unpredictable water flows. In 1971, NSP donated two and a half acres associated with the hydropower facility to the city. The utility also donated the large dam originally used by the Queen Bee and, later, the Byllesby hydropower plant. This critical gift made it possible for the city to expand Falls Park and offer improved public access to the main section of the cascades from both sides of the river.

NSP intended to phase out its coal-burning power plant at the falls location. And three years later, in 1974, the company closed all remaining operations along the cascades and donated its land and buildings to the city. That gift included four acres and a collection of structures needing care or demolition. One giant brick chimney soared two hundred feet. The energy plant was built solidly from concrete, quartzite and brick.

Although the city was pleased to receive the new property, officials realized that possessing title to substantial structures lacking defined purposes posed considerable challenges. "We suddenly owned these buildings," remembered one city official, "and we had to figure out what to do with them."

A twenty-seven-year-old disc jockey and radio talk show host with no political experience surprised many in Sioux Falls when, in 1974, he was elected to serve as the city's mayor. Rick Knobe was a Chicago native who'd graduated from Morningside College in Sioux City, Iowa. Working at Sioux Falls' smallest radio station was his first job. "During my talk radio program, a woman called and suggested I run for mayor," Knobe remembered. "Other people were also urging me to do it."

One of his program's advertisers was RISE leader Roy Nyberg. "Roy told me about the value of Falls Park and the river," recalled Knobe. "I also became friendly with Earl McCart." These men introduced Knobe to Hazel O'Connor. Barely into his term, the young mayor was already familiar with three leading river advocates.

Hazel O'Connor made a lasting impression on the young mayor. "She was old enough to be my grandmother," said Knobe, "but she was ahead of her time, in terms of women's liberation and community activism. She was civil. She didn't cuss. She was feisty in a good way."

RISE now had a stalwart ally at the top of Sioux Falls government. No previous mayor had expressed such an interest in the city's namesake. The man Knobe replaced, M.E. Schirmer, had opposed development of a recreational trail along the river. Knobe proved to be open-minded and unafraid to step outside traditional orthodoxy.

As Rick Knobe began his service as mayor, the environmental movement was spreading its influence across the nation, and the city's new leader was interested in conservation. He promptly created the city's first environmental protection board, and Hazel O'Connor was an obvious appointment. "Rick Knobe," said one city employee, "helped kick-start the environmental push in Sioux Falls' government."

Among Mayor Knobe's new hires was a city planning specialist named Steve Metli, only one year older than the mayor. Metli, a Sioux Falls native, remembered moonlit escapades as a boy navigating dark, mysterious scrapyards so he could night fish at the falls. Later, he spent six years in Germany and observed that the German people admired rather than spoiled their rivers. *Why not honor the Big Sioux?* he mused. The swelling coalition supporting the falls and the river had gained another effective activist in Steve Metli—and this one could contribute to the cause from an especially influential office.

Steve Metli would later say that he and Rick Knobe were the products of a new generation. "As baby boomers, we had different values than our parents had," he said, defending their plan to renovate the river. "People were ready to take that leap." Knobe and Metli both agreed with a RISE proposition that the city acquire as much of the land along the Big Sioux River within city boundaries as possible. This would be the foundation of a greenway corridor, dotted with shoreline parks and connected by a hiking/biking trail system. As the city's chief planner, Metli would lead the charge on this ambitious undertaking. Land trades, property donations, outright purchases and condemnation were all tactics available for use. Metli proved to be a master at solicitations and dealmaking.

One year into his administration, Mayor Knobe unveiled the Sioux River Greenway Plan, an eye-popping blueprint to develop an uninterrupted, twenty-mile-long walking and biking trail following the Big Sioux River within the city limits of Sioux Falls. Improving Falls Park was a principal element in the plan.

The Greenway Plan was a bold proposition, a daring venture. Knobe and Metli were smart enough to comprehend the hazards but fearless enough to charge ahead. This was heady stuff. Much was at stake. In addition to a lengthy walking trail, the city's downtown would benefit from citizen-friendly riverfront development. There would be a charming promenade, cozy amphitheaters, managed and landscaped green space, historic markers. The reward was alluring. A regrettable liability would be converted to a stunning asset. Some felt it was an impossible task, a herculean challenge. Remaking a community blemish into a scenic amenity seemed an insurmountable undertaking.

Private properties were common along the river within city limits, making it impossible to apply visionary governance. Aesthetics and environmental management suffered. Nearly forty pipe outlets discharged pollution and sewage directly into the river. Timbered hideaways harbored piles of refuse. Residents randomly discarded old appliances and furniture on the shoreline. Ragged industrial structures needed to be razed or moved. Flooding was a problem requiring a network of levees and floodways. These features needed landscaping and beautification.

Knobe and his fellow city commissioners adopted the Greenway Plan, committing to transform the city's relationship to the river. Never before had the city so clearly supported an environmental message.

> *Now, therefore, be it resolved by the Board of City Commissioners of the City of Sioux Falls, South Dakota, to preserve the unique natural resource of the Big Sioux River and its tributaries giving due respect to its contributions to our City's heritage, and to make its special features available for the enjoyment of all.*

> *To accomplish this, the Board of Commissioners of Sioux Falls shall follow land use practices which maximize the unique qualities of natural beauty, wildlife habitat and open space along the shores of the Big Sioux River and its tributaries.*

Local conservationists were heartened to witness what was happening in city hall. The newly passed resolution could serve as a constitution of sorts, a long-term guide to Big Sioux River stewardship and supervision.

Knobe and the city prepared a brochure detailing riverside lands to be acquired and distributed more than seven thousand copies to citizens. Before widespread circulation, a ceremonial first copy, autographed by members of the city commission, was presented to Hazel O'Connor.

O'Connor and RISE had played an integral role in formulating the document. In matters regarding Falls Park and the Big Sioux River, O'Connor had risen from outside agitator to inside influencer. Sioux Falls was on the threshold of an historic about-face. No longer would public investment and community interest in the river and Falls Park be an on-and-off matter. Hazel O'Connor was responsible, more than anyone, for that fortunate circumstance.

Chapter 9
PROVIDENCE AND PLACE

Not all ideas associated with the 1975 Greenway Plan progressed from drawing board to actuality. Some rejected suggestions previously appeared in the 1969 concept plan developed by architects at the University of Nebraska. A small marina on the river near Fourteenth Street was proposed. Wisely, that pipe dream was disregarded. Renovation of the old Illinois Central rail depot near Eighth Street and the Big Sioux River as a gateway to Falls Park was planned. That didn't happen. A two-hundred-foot observation tower suggested for Falls Park seemed a bit much. Those 1969 and 1975 blueprints, rejected ideas notwithstanding, encouraged Sioux Falls to think holistically about the park and the greenway and how those outdoor features could be enhanced and integrated to improve quality of life for the entire community.

The city's plan to acquire land along the twenty miles of Big Sioux River that coursed through Sioux Falls involved a variety of properties and structures near Falls Park. Some cases were easily handled. A warehouse near the cascades owned by a local beverage distributor was purchased and razed. Another industrial structure, owned by Land O'Lakes dairy cooperative, was also removed. Property hosting a lumberyard near the park was acquired, and the business was relocated. Dealing with other businesses identified as obstacles required more finesse or aggression.

Each of the affected businesses had their own heartbeat, their own pulse, their own place in the lives of those who worked or did business there. Those enterprises may have blocked the path to a larger and better public park, but their undoing or relocation caused anxiety, inconvenience and disruption.

A truck and auto salvage business, V&O Salvage, occupied five acres alongside the Big Sioux immediately south of the Queen Bee ruins. Rick Knobe reported that the area was covered with discarded vehicles and shabby structures. "The dirt there," he described, "was stained by oil and gas."

V&O's owner, Sam Ogdie, acquired the property in 1964 from his friend Ben Margulies, the fellow who donated land north of the salvage yard, known as the Queen Bee site, to the city. Ogdie's son Sam Jr. recalled visiting his dad's junkyard as a youngster with his brothers. For curious, adventurous boys seeking novel experiences, the setting was a ripe paradise. "We dove off the cliffs of the falls into swirling water that was clear to the bottom," remembered the younger Ogdie. "The shoreline on the east side of the falls wasn't pretty. The old power plant was deserted, and what was left of the mill was falling apart. There were transmission wires crossing the river. Homeless people sometimes slept in the abandoned tunnels beneath the mill, and there was a tramp camp across the road, near the railroad tracks." Sam Jr. recalled at least one destitute who regularly slept in salvage yard vehicles. "He was a former lawyer and had lost a successful career because of alcoholism. My dad would ask my brothers and me to go find Harry in one of the scrapped vehicles and take him out for lunch." On Saturdays in the summer, the Ogdies barbecued at their family business and invited hoboes to join them for a meal.

Sam Jr. and his brothers watched a parade of their dad's friends visit the business for coffee, joke-telling and political discussions. Local businessmen, the country sheriff, judges and even Richard Kneip, a dairy equipment dealer who later served as South Dakota's twenty-fifth governor, socialized there. Workers from the nearby John Morrell meatpacking plant stopped by to relax after a shift. Sam Sr. hosted informal meetings of Alcoholics Anonymous there, and he introduced goats to wander his property, keeping down the weeds.

Sam Ogdie was raised in hardscrabble circumstances. His father, a Syrian immigrant, supported his family as a peddler operating out of a horse-drawn wagon and selling wares like brooms and kettles to customers on the open plains of western South Dakota. In Sioux Falls, as a younger man, Ogdie and his brothers were bootleggers and lawbreakers. Sam was known as a charismatic guy who instigated numerous entanglements with the police. Sam drank hard, partied plenty and frolicked on the edges of society. In 1930, he was convicted of stealing firearms, food items, garment fabric and a child's piggy bank from a Sioux Falls residence and sentenced to six months in the Minnehaha County jail. South Dakota Governor W.J. Bulow pardoned Ogdie, under the condition that Ogdie refrain from becoming intoxicated. The following year, Ogdie was sent to the state penitentiary for five years because he had beaten and robbed a man during a night of heavy drinking. Parole came three years into his sentence, but several months after being set free, he was arrested for breaking into a home. Almost immediately, he was sent back to prison. The *Argus Leader* reported Ogdie had "been arrested so many times police cannot keep count." Another article revealed that during his young life, Sam Ogdie faced arraignment for various charges on more than thirty occasions.

Eventually, Sam Ogdie straightened himself out, opening and operating profitable businesses and watching his children pursue successful careers. Wayne Fanebust's father and Sam Ogdie were friends, and when Wayne completed his military service in the mid-1960s, he returned to Sioux Falls, hoping to buy a fancy car to celebrate. "Sam took me to an auto auction," remembered Fanebust. "I found a red convertible there, a real beauty, and Sam bought it for me and told me he'd set up monthly payments through a local bank that I could afford." Fanebust never forgot that generous gesture. "I loved that car," said Fanebust, "and I loved Sam."

In 1975, a fellow showed up at the salvage yard wanting to sell a truck trailer. Ogdie, then sixty-eight years old, asked for a title, and the man said he did not have one. Ogdie hesitated and then went ahead with the purchase. Soon after, he resold the trailer. As it turned out, the trailer had been stolen, and Ogdie was convicted for receiving stolen property. The court fined Ogdie $250 and suspended a six-month jail sentence. That same year, a much bigger threat to his business and livelihood arose when the City of Sioux Falls declared it must acquire his salvage yard to expand Falls Park. Ogdie refused the city's initial purchase offer. Further negotiations proved futile. That compelled Sioux Falls to exercise its power of eminent domain.

A 1976 editorial in the *Argus Leader* accused the city of "shabby treatment of salvage dealer." Ogdie had been given only ninety days to vacate the location, a place where he had conducted business for a dozen years. Twice, Ogdie tried to relocate the business elsewhere in Sioux Falls. In each case, neighbors near the proposed locations protested. Ogdie supplied a down payment for one of the properties but lost his investment. Finally, at a third location, consisting of fifteen acres in northeast Sioux Falls, the salvage business was reestablished. According to his family, Ogdie received a fair amount of money for his land and moving expenses, and the change in address proved lucrative. "At the peak," said Sam Jr. many years later, "we had over one thousand vehicles at the new site and were the largest truck salvage business in the state." Ogdie ran his increasingly profitable enterprise from that location until 1992, when he sold it to an Iowa-based truck and truck parts retailer.

Richard Kneip, who served as South Dakota's governor from 1971 to 1978, exercised his authority to pardon Ogdie for the stolen property conviction, marking the second time in Ogdie's remarkable life that his record of criminal wrongdoing was cleared by South Dakota's top elected official.

Once Ogdie departed his property near the falls, the city possessed what it needed to more fully develop Falls Park on the eastern side of the river.

The railroad bridge originally constructed by the South Dakota Central in 1906 still blighted the scenery and contributed noise and fumes to Falls Park. Efforts to remove that crossing, a goal continually expressed by a sequence of different mayors and community leaders, would prove frustrating and unproductive. One bridge that was welcomed at the park was a pedestrian bridge, completed in 1977, adjacent to the old NSP powerhouse and situated at a superb vantage for park visitors. "We have a new and thrilling view of the rock formations and the falling water," remarked Hazel O'Connor at the structure's dedication.

NSP's old powerhouse stood, dusty and cheerless, on a stout concrete foundation beside the river. After the utility company gave the historic structure to the city in 1974, Earl McCart and Hazel O'Connor wanted to rehabilitate and convert the building to an undetermined, useful purpose serving the park, but some in city hall weren't so sure. When a bid to tear down the building came in at a whopping $143,000, the viewpoint favoring repurposing the structure gained traction. Determining how to revitalize and utilize the powerhouse would involve debate and planning over the next thirty years.

The quartzite carcass of the Queen Bee also presented unique opportunities. What was left of the mill was eventually placed on the National Register of Historic Places. It was envisioned by park leaders that the building and grounds would add an educational component to Falls Park, presenting a reminder regarding evolving uses of the falls.

Rick Knobe sought reelection in 1979, and Hazel O'Connor became one of his leading campaigners, cochairing a committee that raised money and placed ads for the young mayor that cited his accomplishments, including his open-door style of governing and farsighted leadership on Falls Park improvements and the river greenway. Knobe's first victory had come in a tightly contested runoff against a sitting mayor. This time, he triumphed by an impressive margin. In a five-person field, the incumbent attracted 64 percent of the vote, more than tripling the total of his nearest competitor. Obviously, citizens liked what thirty-two-year-old Rick Knobe was doing to improve their community.

In 1981, the *Argus Leader* reported that a significant portion of the original greenway network had been completed. That same year, a bronze plaque honoring Hazel O'Connor was placed on an overlook at Falls Park. She later told a reporter she was enthused about the greenway, the river and the park. "The city has done well," she said. "But you can't let them lag. If you don't jack them up once and a while and let them know about the history, people get comfortable."

Four years later, the city lost its foremost Falls Park activist. Hazel O'Connor's death was mourned by family, friends and many acquaintances, and their recollections about a person who made a difference covered a wide array of themes and subjects. Former park board president Peder Ecker expressed his gratitude. "Credit for Falls Park," said Ecker, "goes to Mrs. Hazel O'Connor and her constant prodding and interest in the renewal of the park."

Hazel O'Connor's friend Carol Mashek explained that a betrayal by city officials in 1966 had had a lasting impact on O'Connor. O'Connor had led efforts to preserve a historic home built in the late 1800s by city pioneer Dr. Josiah Phillips and his wife within present-day Terrace Park, Mashek recalled. That once majestic structure, built from local quartzite, had fallen into disrepair, and park board members wanted it torn down. O'Connor and a group of historic preservationists gathered hundreds of signatures on a petition urging the home's renovation and public use. They were told by parks officials that the petitions would be considered at a park board meeting to be held on Tuesday, October 4, 1966. But during the preceding Saturday morning, the structure was demolished. Hazel struggled to trust city hall after that, Mashek said.

O'Connor served on the RISE board of directors until her death. Replacing her was her daughter-in-law, Barbara O'Connor. "I called her Mom," said Barbara, "and she taught me a lot, including a love of volunteering in the community." Barbara herself became an involved citizen, serving meaningful causes and opening a popular shop on Phillips Avenue named Prairie Star that promoted Native American art and culture. Her business neighbor, Jeff Danz, founder and owner of the Zanbroz Variety store, described her as among a small group of downtown's ablest and most engaged defenders. "She was a pillar of our efforts to reinvigorate downtown," said Danz.

In 1987, the city released an updated report on the greenway and riverfront, including components of Falls Park. "Natural areas remaining in Sioux Falls," said the report, "will be protected so that future generations will always have the opportunity to experience, appreciate and learn from the gentle splendor found in them."

Sioux Falls mayor Jack White and the city commission were all in on pushing ahead with the next phase of the greenway. Several years earlier, the city had created a citizen-run Greenway Task Force to help formulate and advance the plan. The task force was elated at the progress completed during the previous decade. "Most of the physical pieces are in place," task force members agreed, "to create a greenway system comparable to any in the nation, especially for a city the size of Sioux Falls."

Continuing to realize the city's vision would require a greater investment of the city's own money. Sioux Falls had first undertaken the greenway project and Falls Park enhancements at a time when federal assistance to aid community projects was readily available and simpler to secure. Local taxes would be needed to fund future park and greenway projects. "During the 1970s," explained the 1987 report, "federal monies from a variety of departments played an important role; however, this role has diminished over the years with changes in the national economy and alteration in the political philosophy and priorities of succeeding administrations."

Rick Knobe acknowledged the critical role the federal government played in the development of Falls Park and the greenway system. "We learned to work with the federal government and got good at finding money there," Knobe reported. Federal assistance was used for assorted purposes, from surfacing trails and landscape beautification to private property acquisition. It is fair to say that without the generosity of the federal government, the city's greenway system and Falls Park would be less accessible, less successful and smaller in scale.

In 1991, Mayor White, Steve Metli and his assistant Mike Cooper decided that a comprehensive master plan for Falls Park, prepared by professional landscape architects, was overdue. Until then, city officials had mostly implemented their own ideas for the modest park improvements that had been undertaken. To pursue more dramatic and meaningful upgrades to advance the park, the city hired Big Muddy Workshop, a firm based in Omaha, Nebraska. Cooper, who would eventually manage both the city's planning office and the parks and recreation department during his public service career, would be the city's project manager working directly with Big Muddy.

The owner and leader of Big Muddy was John Royster, a landscape architect who grew up in Canistota, South Dakota, thirty miles from Sioux Falls. Royster's feelings for the falls were rooted in his childhood, when he and his family occasionally drove from their home to picnic at the park. As an expert in designing and directing development of outdoor spaces and parks, Royster brought a thoughtful and sensitive creativity to his work, explaining that his intention was to "celebrate and conserve a landscape." Big Muddy Workshop, he said, specialized in "environmentally responsible designs that encouraged people to encounter natural landscapes so they can form positive relationships with the land."

Both locals and out-of-towners were drawn to the place, Royster realized, to savor the sensational scenery, but he felt there was more to experience. Royster began to study the park, its natural conditions and the buildings and roadways surrounding it. It was apparent that the park remained oriented to visitors viewing the falls from their cars or trucks. "It was our objective to get people out of their cars," said Royster. "We wanted people to become pedestrians at the park and view the falls in ways they'd never seen." Mike Cooper remembered citizen resistance when it was announced that walking, rather than riding, would be the park's new emphasis. "It was," acknowledged Cooper, "a big change, but it needed to happen." Royster's wife, Katie Blesener, vice president of Big Muddy, explained that their plan was to connect visitors to nature. "We wanted people to have an authentic rather than a commercial experience," she said.

On a summer afternoon in 2024, John and Katie stood together in the shade of an oak tree at Falls Park describing the urban nature sanctuary they helped design. A massive rainstorm had rumbled through the area earlier in the week, and the falls were roaring. A steady parade of visitors strolled past, and all eyes were trained on the stunning spectacle. The more adventurous wandered about on walkable quartzite outcrops nearer the raging river. Phones held to faces were common as photos and videos of falling water were captured. A man from Alabama snapped a few photos and said, simply, "Never expected to see this here."

"It is gratifying," John Royster declared, "to witness so many people admiring this special place."

Three decades after Big Muddy and Sioux Falls transformed a raw park brimming with potential into an inspiring and appealing destination, Royster pointed out the physical elements incorporated into the master plan. The multiyear process began with the basics. "When we first examined the park property," recalled Royster, "it was overgrown with invasive species, and weeds and bushes sprouted from gaps in the quartzite along the river. All that needed to go." There was patchy grass and trees that needed tending. Some trees were removed, and new trees were planted. Purposeful landscaping was applied. The perimeter of the park, affecting the ambiance of the park's interior, had been improved by removing industrial enterprises, but more relocations needed to happen. Homeless people continued to build makeshift shelters in the park's thickets and woods, a fact that saddened Royster. But he knew that situation must change.

Parking lots crowding the river were bulldozed and replaced by better-built versions set back from the falls. Roadways winding toward those parking lots encouraged slow travel. New sidewalks offered breathtaking vistas and led visitors to appealing viewing platforms along the river. Visible quartzite formations situated throughout the park were integrated into the park's features as environmental highlights. Overhead power lines were buried. Historically themed streetlights were added alongside roadways and walkways. The city's horse barn, a curious relic from yesteryear located in the northeast corner of the park, which once served as a storage building for the city's utility department, was remodeled and opened as an art center.

Details such as the color of concrete were considered. "Steve Metli came up with the idea of tinting some concrete sidewalks reddish, to complement the quartzite," Royster explained. Katie Blesener wrote content for new educational signage.

To determine locations for six new concrete viewing platforms, Royster made numerous site visits and slowly walked and carefully studied the park, the cascades and the shoreline. He reviewed topographic surveys, paying special attention to rock formations offering advantageous views and solid footings to support construction. He considered factors such as height and orientation.

South of the railroad bridge, on the west side of the river, twin viewing platforms—one low, one higher—were constructed to offer different views of the uppermost stretch of the river's cascades, a broad, shallow channel with lots of visible, low-slung quartzite. In this stretch of river, high flows don't rumble; they riffle over sheets of smooth rock. This is where the cascades commence, slightly upriver from the most dramatic stretches of the falls.

Near the twin overlooks, Royster stood on what used to be Seney Island and noted that there was no evidence the island ever existed. A short distance upstream rose enormous buildings associated with a new real estate development along the river called the Steel District. Royster had situated the two overlooks at the southwest corner of Falls Park so a viewer could turn away from the river and admire a pair of architectural icons poking through the Sioux Falls skyline: the clock tower of the old county courthouse and the twin spires of the Cathedral of St. Joseph. Other Falls Park overlooks had also been situated to permit viewers to observe those two distinct buildings and others in the downtown district. That panorama was now gone, obscured by more recent structures. The long-distance viewshed that formerly connected the park to a historic cityscape and a vast dome of sky stretching to a distant horizon had met the same fate as Seney Island. Royster regretted the dramatic change. "I think the viewshed was ruined," he said as he stood in the sunshine, gazing southward.

Farther downstream, north of the rail trestle, two overlooks on the west side of the falls were added, each offering spectacular views of crashing water. These viewing platforms were instantly inviting to park visitors as they walked from a nearby parking lot.

Two new observation decks on the east side of the river also presented breathtaking and varied exposures to the falls. One platform, on a rocky bank below the Queen Bee mill, replaced the park's first viewing platform, a deck that had been assembled on top of the stone portion (the lower portion) of the Queen Bee's turbine house after the structure's upper section, made from wood, was removed. This feature had been Earl McCart's idea nearly thirty years earlier. John Royster convinced city officials to eliminate the turbine house deck and return the small historic building to its original appearance. "The view from the park's earliest overlook," Royster recalled, "failed to

Falls Park and cascades before the Steel District real estate development changed the panorama. *Courtesy of South Dakota Tourism.*

capture the power and scale of the falls. It was too high and situated at an angle that made the falls look less significant."

Royster's new platform was located closer to the river and delivered an intimate view facing tumbling, turbulent water. "I'm about five foot seven," said Royster, "and when I stand on that platform, I am looking point blank at the precise place where the river drops over a long, steep precipice." It is a startling sightline, a closeup confrontation with the gushing vein of a channel that abruptly bends straight down and sprays a mist over park visitors standing there. "This," Katie Blesener declared, "is my favorite place in Falls Park."

Viewers using the second new platform on the eastern side of the river found themselves positioned alongside and slightly above the falls, able to peer downward from that raised perspective at the back side of a cataract's crest. "I selected the location of each viewing platform to provide the viewer with a different perspective of the falls, so they could watch water doing something they could not see elsewhere in the park," explained Royster, pointing to flows stilled in a small, composed pool before they plummeted like an avalanche down a sheer vertical drop.

An observation tower was also proposed, and city officials suggested it be located alongside the most dramatic stretch of falls. John Royster preferred that the tower be connected to a new visitor center that would be situated farther away from the river, and his recommendation was approved.

Mike Cooper described the process of deciding the height of the tower's elevated viewing balcony. That verdict, explained Cooper, was not guided by engineering computations or architectural calculations. "We used a tree-pruning boom to survey our options." At a predetermined time, the three men, Cooper, Royster and Steve Metli, as well as a city forester to chaperone the examination, arrived at the park to solve the dilemma. The day they chose was breezy, an intimidating factor considering the airborne instability of the lightweight lift. Each of the three men, accompanied by the forester, rode separately in the apparatus to judge various heights and perspectives. "John and I were nervous about rising too high in the strong gusts," Cooper explained, "but Steve wanted to go higher and higher." Finally, as the lift wobbled in the wind and reached its limit, Metli was satisfied. "He wanted that platform built as tall as the lift could go," grinned Cooper. And it was: a hard-earned fifty feet.

Another challenge was the railroad bridge spanning the falls within the park. That river crossing, originally opened in 1906 and upgraded in 1947, obstructed the vista looking south over the cascades. Many wondered why

A Natural & Cultural History

Top: A pedestrian bridge and a strategically located overlook at Falls Park. *Courtesy of South Dakota Tourism.*

Bottom: John Royster's Falls Park design emphasized foot travel and intimate views of the cascades. *Courtesy of South Dakota Tourism.*

this conspicuous and unsightly feature couldn't be moved, like the city had evicted other businesses encroaching on the park. Local railroad historian Ed Monson explained that railroads enjoy a level of untouchability. Once their tracks are laid, said Monson, it is practically impossible for a community to force a railroad to relocate. "You can negotiate with a railroad," Monson explained, "but you cannot dictate to them."

Park advocates grew dejected and weary of the debate, and the bridge was never removed, but Steve Metli managed to score one train-related victory. Metli was upset because the railroad had begun parking idle train cars on its bridge, adding to an eyesore that already diminished the park. "Steve began discussions with the railroad," recalled John Royster, "and suddenly the practice of parking train cars on the bridge was discontinued."

The city flirted with razing the NSP building as late as 1987. Dangerous amounts of asbestos were discovered in the structure, and the cost of removing this material approached a quarter-million dollars. Pigeons and other birds had invaded the abandoned building, and thick layers of bird waste adhered like hardened glue to the walls and floor. But John Royster and Mike Cooper viewed the building as an asset and successfully promoted rehabilitation. Once it had been repaired and cleaned, the exterior presented a charming façade. The renovated interior was spacious, with surprising architectural details, and the building's west end offered access to a panoramic prospect of the falls.

The city would eventually spend more money than anticipated to refurbish the 4,200-square-foot structure, and not until 2004 would there be a public use for the historic building. That's when Mike Cooper announced that a new business, named Falls Overlook Café, would use the entire main floor to serve guests in a family-style restaurant. Over subsequent years, other vendors and eating establishments would try their hand at making the facility turn a profit. The name of the restaurant would stay the same, but the approach to service and food offerings would change.

Much of the physical work that improved the park would happen during the mayoral tenure of Gary Hanson, from 1994 to 2002. "Falls Park was one of my top concerns as mayor," Hanson explained. He remembered dumping garbage at the city's landfill near the falls when he was young. He also remembered swimming as a youngster with friends in an abandoned quarry a short distance east of the park. "Considering everything that was dumped into that pond, it's remarkable we survived," said Hanson. "I was sick every so often, with red rashes on my skin and discolored swimsuits."

A Natural & Cultural History

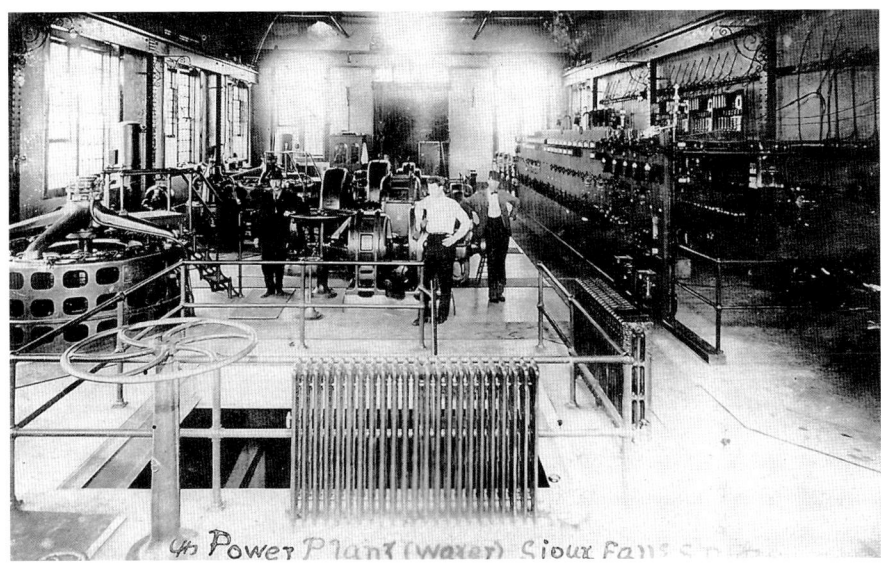

Top: Original interior of the Sioux Falls Light and Power operations center, now the Falls Overlook Café. *Courtesy of Ed Monson Collection.*

Bottom: The reconditioned, repurposed Northern States Power building became a central feature of Falls Park. *Courtesy of South Dakota Tourism.*

Landscape cleanup and restoration at the Northern States Power site involved removing smokestacks, cooling towers and storage facilities. *Courtesy of South Dakota Tourism.*

Purifying that polluted pond and extending Falls Park east and north to better serve the nearby Whittier neighborhood became one of Hanson's unfulfilled dreams. He initiated discussions with officials at the John Morrell meatpacking plant, located a short distance downriver from the park, hoping the city could acquire the Morrell site and move the slaughterhouse to an out-of-town location. "I hoped we could transform the site of the meatpacking plant, a place I considered the least attractive part of Sioux Falls, into one of the most attractive parts of the city," Hanson said. Unfortunately, that ambitious idea became another unrealized aspiration. Hanson also tried to resuscitate Seney Island. "We did soil borings," he explained, "and discovered the site was filled with contamination and garbage. The federal Environmental Protection Agency told us to cap the area with clay, and that's what we did."

Despite the setbacks, Hanson and his administration achieved notable progress at Falls Park. Developing Big Muddy's design was a Hanson priority, and it would prove to be a spendy proposition. It is estimated the city spent over $8 million on Falls Park during this phase of park improvement work. Providing better visitor access to the revamped park became the next major challenge facing the city.

"One of the problems facing Falls Park at that time," explained John Royster, "was despite the relative proximity of the falls to downtown, there was no easy way to get to the falls from downtown." Walkers and bicyclists were unable to follow the river from downtown to the falls. No greenway or trail existed there. Drivers coming from downtown or the south were forced to follow an awkward route, dodging industry and following streets lacking specific directions to Falls Park.

The team at Big Muddy provided early design ideas for a concept known as Phillips Avenue to the Falls, an initiative unveiled in 1987 by Steve Metli. The plan would extend Phillips Avenue (which, at the time, dead-ended at Fifth Street) from downtown directly to the park's entrance. It was a brave and far-reaching dream by Sioux Falls city government, and in the way stood a swath of private property.

Lengthening Phillips Avenue northward to the park would create only four new blocks of street, but the plan's impact would be gigantic for Sioux Falls. Indeed, Steve Metli's overarching goal was to completely remake a sizeable section of the city's center, replacing industry with residential, retail and green spaces. An unattractive, noisy expanse of dust and grime would be exchanged for an appealing, livable neighborhood.

Steve Metli and his boss understood that the ambitious plan would necessitate evicting several local property owners and removing operational railroad infrastructure. These actions, they knew, would be contentious.

Those train tracks and switching yard requiring relocation were part of the same rail features vigorously opposed by Richard Pettigrew and others in the 1920s. Pettigrew had warned that it was a short-sighted decision by the city to allow the Chicago, Milwaukee and St. Paul Railroad to impose an expanded trainscape on an area near the falls. But the railroad warned city leaders that without new rail facilities, Sioux Falls would slip behind Sioux City as a regional warehousing and distribution center. Pettigrew—ironically, the city's most important railroad promoter—was dismissed in that instance as an anti-progress sentimentalist, and the railroad prevailed.

Those familiar with Sioux Falls may know that the railroad expansion approved by the city in 1926 included a bridge that spanned the Big Sioux River a short distance south of Sixth Street, near the current location of the Arc of Dreams sculpture, and tracked north, crossing Phillips Avenue at Fifth Street and blocking Phillips at that point. The dead-end would disappear under Metli's Phillips to the Falls plan. So would the large railroad switching yard.

Another significant obstacle was a scrap metal and recycling enterprise located north of where railroad tracks crossed Phillips Avenue, between Fourth and Fifth Streets. Pitts Inc. had conducted business from this address since 1920 and owned about sixteen acres that were critical to the city's plan.

Near Pitts was a parcel owned by a family who formerly ran a lumberyard on the site. A warehouse there was rented to a building supply firm named I-29 Brick. It may have been a small piece of land, not quite two acres, but like the Pitts property, was vital for the city's Phillips Avenue extension.

Pitts was a highly industrialized setting, a typical junkyard, with scrap metal piled high across dirt and cinder lots. Long-necked cranes moved product to trucks and trains, and several shabby structures were scattered about the large, dusty property. It was the last of three large salvage yards that had once surrounded Falls Park.

Negotiations between the city and Pitts had long been unsuccessful when a massive fire in August 1999 destroyed much of the salvage yard, including the company's office and a warehouse. This was the second large fire at Pitts in seven years. In 1992, an old wooden railroad roundhouse located on property acquired by Pitts, built in 1928 and later abandoned, had burned down on a chilly November day after at least one homeless vagrant lost control of a fire intended to provide warmth. Officials determined that the 1999 fire was intentionally set, but an arsonist was never identified. As with the Queen Bee fire some forty years before, much of what was lost was uninsured. Nevertheless, Pitts decided to stay in business, retaining its fifteen employees. Gary Hanson had grown frustrated with the pace of progress, so he hired a professional negotiator to work out the details. That negotiator also dealt with the owners of the warehouse serving I-29 Brick.

In May 2001, the impasse with Pitts was resolved when the company agreed to vacate the property within eleven months. The cost to Sioux Falls was $1.6 million, and Pitts's managing partner claimed this was an inadequate sum to relocate and reopen his company.

Pitts would leave behind a site contaminated by lead and petroleum and needing substantial environmental remediation. Soil borings in the Pitts and old Seney Island vicinity revealed toxic levels of coal tar that was likely dumped there by a nearby coal gasification plant. Contaminated earth was stripped away, and the area was capped with clay. To perform this substantial and necessary cleanup, the city was aided by federal Environmental Protection Agency expertise and money.

The city had just commenced eminent domain proceedings against I-29 Brick in early 2003 when a settlement was reached, ending a seven-year stalemate.

Extensive consultations with the Burlington Northern Santa Fe Rail Company (BNSF) regarding tracks, trestles, crossings and the switching yard relocation began during Gary Hanson's first term as mayor and would prove to be as complex and confounding as anticipated. "My staff and I negotiated with the railroad over different options," Hanson said, "but they were intransigent."

Mayor Hanson attended a meeting bringing together various rail interests that was held in South Dakota Governor Bill Janklow's conference room. The atmosphere in the room, recalled Hanson, was tense as the governor pressured rail executives. He reminded them that their trains used tracks owned by the State of South Dakota and that by not cooperating with Sioux Falls, they jeopardized that arrangement. "The governor really drilled railroad officials, and they did not put up any resistance," said Hanson. "However, after the meeting, they made few concessions."

Under Hanson's successor, Dave Munson, citywide impatience with the railroad intensified. "I felt we had talked about it long enough," said the new mayor. Munson reached out to the president of the railroad and sensed an opening. He pushed his staff and others to compel a satisfactory resolution of the situation. "We began getting an endless stream of phone calls and letters," said a BNSF spokesman. "There was a sense of greater urgency."

The railroad finally agreed to relocate some of its operations, including the switching yard, to a different section of the city. BNSF also reinforced and upgraded the underpass of its crossing over what would become a stretch of Phillips Avenue located near the entrance to Falls Park. Auto traffic approaching Falls Park from the south or leaving the park and headed toward downtown navigates an S curve on Phillips Avenue that passes beneath the improved bridge.

On October 30, 2004, the sparkling, four-block stretch of new street split by a landscaped boulevard that fulfilled the connection between downtown and Falls Park was dedicated during a rousing celebration held at Falls Park. In the annals of Sioux Falls history, completing that link was a momentous event. "The city has come to appreciate what makes Sioux Falls Sioux Falls," declared Bill Hoskins, director of the city's Siouxland Heritage Museum.

City leaders understood that more than a new street was being feted. A wide corridor bordering the freshly opened throughfare posed both public and private sector opportunities.

At the dedication, Dave Munson acknowledged the contributions of his predecessors, city employees and officials representing state and federal government. "This has been a collective effort by a lot of people," he proclaimed, before predicting the outcome of the accomplishment. "We'll [one day] reflect on the fact that we saw the beginning of a transformation of this part of the downtown area."

"Sioux Falls mayors during three decades had advanced Phillips to the Falls," Gary Hanson later said. "I credit Dave Munson for moving the project across the finish line."

"There were a lot of skeptics," Steve Metli remembered. "Everyone knew it would be a long process. No one knew it would be seventeen years."

Metli predicted that connecting downtown to Falls Park would especially benefit the park. "We don't ever want to see Falls Park fall into disrepair," he said. "With Phillips to the Falls, downtown will always babysit and oversee the Falls."

Project costs were substantial. Millions had been spent to acquire property. Outlay for street construction was about $4 million. Pollution remediation required at least $1 million, and more contamination would later be uncovered and dealt with. One estimate put the project's initial price tag at $9 million, not counting costs to relocate railroad facilities. That aspect of the project, railroad-related expenses, exceeded $40 million.

Three years after Munson's ribbon cutting, an alliance of developers, the so-called Uptown at Falls Park group, purchased nearly five acres of land adjacent to Phillips Avenue from the city for $2.6 million and unveiled an expansive design to be built over ten years. Land formerly occupied by train tracks, junkyards and shabby industrial buildings would instead accommodate shiny modern residential and commercial spaces. There would be a pedestrian plaza and a twelve-story hotel. One building would partially span a street. An entertainment district would make this new neighborhood a genuine destination. Steve Metli's vision would be realized.

Nearby historic industrial structures were remodeled, adding residential lofts and commercial frontage along Phillips Avenue and Main Street. This effort was led by local architect Jeff Hazard. Hazard transformed an old seed warehouse into a modern gem with condominiums on the top floors and micro-retailers at street level. For a conservative community like Sioux Falls, this new type of development was genuinely innovative. City leaders tagged the district Uptown, a moniker lifted from a Minneapolis locality.

Craig Lloyd, a Sioux Falls–based developer, led the Uptown group and also pursued his own projects along and near Phillips Avenue. Lloyd and his wife had built a vast and diverse portfolio of real estate holdings in Sioux Falls, including significant buildings along the Big Sioux River in downtown. Lloyd and Steve Metli had become best friends and often took bag lunches to Falls Park, where they walked the grounds to pick up litter.

By 2011, Lloyd's vision for Phillips Avenue had been downsized, but the newer version was still impressive. Two hundred residential units in two expansive multiuse buildings, called the Cascade at Falls Park, remained part of his plan. Commercial space had been reduced, but ground-level retail and entertainment options occupied the upscale apartment buildings he built along Phillips Avenue.

Across the street from Lloyd's project was city property designated Falls Park West. Instead of permitting commercial development on this piece of land, the city agreed to cooperate with the national mission of a live music promoter called the Levitt Foundation. A local nonprofit group called Levitt at the Falls was established to help raise funds to build a spacious amphitheater and stage, and the City of Sioux Falls agreed to maintain the property. The local nonprofit also assumed fundraising for the majority of expenses associated with presenting fifty free concerts each summer; some contributions to that cause came from the Levitt Foundation. Visitors to the amphitheater noticed the band shell's wavy roof. That design saluted the nearby river and falls.

Another sizeable industrial site located between downtown and the falls was owned and used by local company Sioux Steel, Hazel O'Connor's longtime nemesis. The business announced it would leave its headquarters and manufacturing facility after a century of occupation, permitting redevelopment on this choice piece of land along the Big Sioux River and the uppermost section of the cascades. Lacking landscaping and any sense of aesthetics, Sioux Steel's property, if not transformed and upgraded, would have blighted the otherwise improving neighborhood.

Craig Lloyd's proposal to repurpose Sioux Steel's property impressed the Rysdon family, owners of Sioux Steel, and they sold him their prized real estate. Lloyd's new development would include an upscale hotel and convention center, 140 loft-style apartments, 20 high-end condominiums and a nine-story office building that sat over part of what had been Seney Island. Shops, bars and eateries would operate at street level. To honor the property's longtime owners, Lloyd named his new development the Steel District.

Much of the Steel District would front or overlook the river and a narrow greenway highlighted by a wide walkway. That walking-biking trail, stretching 1,100 feet along the Big Sioux River within The Steel District, would blend seamlessly into the citywide trail system. "You won't really know where the public and private spaces delineate," a Lloyd executive said, "because our goal is that it will feel very open to the public."

On July 20, 2021, one month before the Steel District's groundbreaking, the Sioux Falls City Council voted to rename the neglected Kiwanis Park property to Lloyd Landing. An agreement had been negotiated by the city with Lloyd Companies, establishing the next official name for a slender slice of land with a controversial history.

Lloyd Landing would occupy some of the same shoreline property that Sioux Steel had illegally created in the mid-1960s. That nonnatural two-acre parcel had been seized from Sioux Steel by the State of South Dakota and turned into a city park. Sioux Steel's incursion happened near where the channel bordering Seney Island's western side diverged from the Big Sioux's main channel. That western channel had been filled by garbage and other materials, and its old route was now occupied by Steel District development.

The newest use of Sioux Steel's unlawfully fabricated property would feature the boardwalk, benches and chairs on tiered river overlooks, loitering spaces of natural grass and artificial turf, a small dog park and a picnic shelter. After her clashes with Sioux Steel, Hazel O'Connor might have been pleased at this convoluted outcome. Sioux Falls residents and visitors would be welcomed back to enjoy what remained of this scenic geography.

On the eastern side of the river, not far south from the falls, local businessman Jeff Scherschligt acquired and razed an old feed mill, the Zip Feeds building, once known as the tallest building in South Dakota. On that property, Scherschligt created a striking structure he called Cherapa Place. After acquiring four adjacent acres from the City of Sioux Falls, Scherschligt added three buildings to his Cherapa development. Like Lloyd's Steel District, Scherschligt's real estate project followed the Big Sioux River downriver from downtown toward Falls Park. His attractive new neighborhood offered over 300,000 square feet of office and retail space, 234 apartments, 11 condominiums and 1,100 parking spaces. Fronting Scherschligt's buildings and bordering the river was an extension of the city's nicely decorated riverside promenade. Patches of prairie, perennial flowers and showy Sioux Quartzite slabs were incorporated into the landscape.

There was one significant impediment to constructing substantial buildings along the river in the vicinity of the falls. Beneath a thin layer of soil was a fat band of bedrock. Jeff Scherschligt's contractors spent four months blasting through Sioux Quartzite to gain the necessary space for foundations and parking. "We went down about fifteen feet to add underground parking areas," said Scherschligt.

Scherschligt's personal residence, situated on the northwest corner of the BanCorp Building's tenth floor, commands a breathtaking view of Sioux Falls and Falls Park. The Sioux Falls native excitedly pointed west to the old courthouse tower and the twin spires of St. Joseph Cathedral, two prominent architectural features John Royster had intentionally incorporated into the Falls Park viewshed, and then to the silvery blue river flowing northward to the falls. "That's where Seney Island used to be," said Scherschligt, gesturing toward Falls Park West and the Steel District. "Too bad we lost that."

But Scherschligt applauded the changing environment. "Reshaping the riverfront represents the next century or more of Sioux Falls life," he predicted. "Sioux Falls has long dreamed of creating a San Antonio–type concept along our riverfront with restaurants, bars and living spaces bringing our citizens together."

According to Scherschligt, concentrated developments like Cherapa are driven by how some people want to enjoy their lives. "There is an allure to places like this," he explained, referencing the river and its riparian corridor, "and density is a goal of urban planners as it is an environmentally friendly approach to creating neighborhoods within a city. Folks are realizing that urban sprawl, traffic and lengthening commutes cause increased costs and frustration. Residing in a dense urban environment is in style, and developers are reacting to what people want."

No doubt, Sioux Falls' population and economy were booming, and the city's inevitable trajectory included suburban sprawl and traffic congestion. Preserving and expanding outdoor experiences for Sioux Falls residents was an oft-stated objective of city planners and leaders. Walkability within the city was a newly emerging factor in determining quality of life. Neighborhoods such as the Steel District and Cherapa, located near or including commercial districts, offered the least stressful, carbon-free commute option. Sioux Falls' newest riverside developments were also stylish and metropolitan.

Expanding the city's trail system and greenway on the west side of the river along the Steel District and controlling the river in that area would cost about $16 million, with one-third of that money used to rebuild the old dam that had served the Queen Bee mill before being lengthened and fortified in 1907 by Henry Byllesby for his hydropower business.

Don Kearney, director of the Sioux Falls Parks and Recreation Department, explained the primary purpose of the reestablished dam. "We need the dam to raise water levels so that when you look at the river, you're not just looking at a small stream. [The river behind the dam] will be a pool of water...allowing for people to canoe or fish and just really have an outdoor experience downtown."

Construction on these city projects followed the onset of the COVID-19 pandemic in the United States in 2020, after which parts of the U.S. economy floundered. Projects like the Steel District were delayed. A federal aid program, the American Rescue Act, targeting community development, job creation and quality of life enhancements funded $9.5 million in costs associated with the Steel District greenway, including all dam-related expenses. The City of Sioux Falls was able to fund other project costs. Despite challenges brought by the pandemic, Sioux Falls' general economic trendline continued to escalate. When Rick Knobe served as mayor, in 1980, the city's annual budget totaled $44 million. Forty years later, that number approached $600 million.

Early concepts associated with a rewrite of Falls Park's master plan were made available to the public in 2022. This redesign was awarded to Confluence, a landscape architecture and urban planning firm, and was led by Jon Jacobson, senior vice president and principal in the company's Sioux Falls office. Confluence also designed the trail and greenway along the Big Sioux River adjacent to the Steel District.

According to Confluence and city officials, park changes would include significant additions, with an emphasis on active recreation and expanded amenities. The older section of Falls Park, nearer the actual cascades, would remain the chief attraction. Increasing access to the park from the east and expanding the park in that direction were proposed features. So was cleanup of the old quarry pond where Gary Hanson dared to swim as a boy. Adventurous youngsters in the future could enter the pond without worry or possibly soar above it on zip lines. A pavilion for large gatherings would be added to the central portion of the park. A one-hundred-foot viewing tower would soar above the falls. Vehicle corridors would be realigned, opening up improved sidewalk circulation within and near the park. Better parking and a new pedestrian bridge and visitors center were other highlights. Confluence's long-term plan intended to relocate a rail line along the eastern side of the park and facilitate private sector opportunities for that area, including housing and commercial development. Finally, imagined city leaders, urban renewal could happen on the west side of the Whittier neighborhood.

North of the Steel District and the Levitt amphitheater, on the western side of the river, was Jacobson Plaza, an undeveloped section of Falls Park West that would be designed by Confluence. Here there would be a recreation smorgasbord, including a refrigerated ice-skating loop, a splash pad, an enormous playground for children and a large dog park.

Jon Jacobson (no relation to the plaza's Jacobson family), described his philosophical approach to expanding and modernizing Falls Park. "Because Falls Park is surrounded by a city and is located downtown," said Jacobson, "it is going to be impacted by the urban densification surrounding it, and the park will have an urban aesthetic. Modern human society typically prefers not to have a piece of 'untamed nature' out their front door in an urban downtown. This is different than the typical state park that is located in a rural area and offers a 'nature'-type aesthetic, with less visual impact by humans."

Updating the master plan for Falls Park was especially challenging, added Jacobson, "because Falls Park is a complex park that is simultaneously a neighborhood park, community park and destination park all in one."

Jacobson's opinions about park development recalled comments made by city parks leader John Browning in the early 1960s. Browning advised a group of park activists, including Hazel O'Connor, that the community would one day need to determine what sort of park it wanted: modern or rustic. Jon Jacobson preferred modern, and that's what city officials wanted as well.

Several noteworthy physical changes to Falls Park were completed by the city before Confluence and the city revealed their proposals. Two new viewing terraces were added to the park in 2019. Construction of the railed concrete platforms eliminated walking access to sections of smooth quartzite where park visitors could stand beside rushing water. Those perilous areas became slippery when wetted by mist or rain. Drownings had become a serious issue at the park, and the problem was especially acute when high, raucous flows drew curious onlookers closer than they should be. Between 1980 and 2018, mishaps claimed at least ten people, and many rescues took place. Warning signs were posted to discourage people from straying onto precarious shorelines. Tragedies were dramatically reduced. Swimming and boating in the park had been banned by 1990.

It became a springtime ritual for Sioux Falls residents and sightseers to visit the cascades of the Big Sioux River when high water rumbled at deafening volume. Those who witnessed this impressive spectacle also observed river flows churning mounds of brown, bubbling foam and delivering a foul stench. At the falls, there was always good scenery, and sometimes there was bad water.

High water once drew the daring to swim and dive at the falls. Both activities are now prohibited. *From the Sioux Falls* Argus Leader.

Agriculture and industry had intensified in the Big Sioux basin, and Sioux Falls had grown rapidly, increasing the city's impact on the river. Those who drowned in the falls likely ingested plenty of pollutants.

Since cities were first settled, issues related to the deposition of unwanted waste generated by commerce and homelife have confounded citizens. Convenience, complacency and inadequate technologies dictated solutions. Downriver or down a hole went discarded and disregarded rubbish and pollution. That was what happened at Sioux Falls.

Rome and Constantinople had the world's first sewer networks, designed to collect and carry human waste to a nearby river or sea. London created a sewage system in the mid-1800s, in response to disease outbreaks and offensive odors carried by the sewage-laden Thames River.

As the population of the United States grew, so too did the need to manage sewage and other forms of pollution. Early on, the same approach used in Rome was utilized in American communities: open ditches carried waste to nearby rivers, lakes or the ocean. Dilution of pollution was the principal solution. By 1900, buried sewer pipes were in use, but they emptied into the same waterways where the surface ditches they replaced had discharged. The evolution of sewage treatment was highlighted by the nation's first chlorination plant, opened in 1914. Imperfect as those pioneering treatment facilities were, they foretold a new era and attitude about waste. Open-air sewer canals and contaminated lagoons were being relegated to a regrettable past.

Much of the impetus rallying the nation to address water quality came from the Izaak Walton League, a national nonprofit conservation organization founded in 1922 and named after an English writer whose passions were nature and the outdoors. Izaak Walton published the legendary book *The Compleat Angler* in 1653 at the age of sixty. The organization honoring Walton was fondly nicknamed the Ikes.

In Sioux Falls, the local Ikes chapter mounted a convincing campaign in 1925 promoting a new sewage treatment facility. A delegation of chapter representatives and farmers and residents living downstream from Sioux Falls appeared before the Sioux Falls city commission and described the Big Sioux: "The river is milky in color and is covered with a thick, slimy scum. The odor is so offensive that it is sickening to human beings and hangs in foggy clouds over the river along the valley." The city, implored a worried resident, needed a "scientific" method to address sewage issues. It was revealed that during the month of June 1925, a daily average of 3.6 million gallons of raw sewage was expelled from Sioux Falls into the Big Sioux River.

"You cannot dump four million gallons of sewage into a stream like the Big Sioux River and expect the river to purify itself," warned an Ikes leader.

To remedy the crisis, the city proposed routing several major sewage lines to a new treatment facility, to be built downriver from the falls. It would be an expensive endeavor, over half a million dollars, and a public vote on the matter was announced. The local Ikes opened a campaign office in Sioux Falls to promote the treatment plant, and proponents prevailed in a lopsided demonstration of concern, with 4,095 voters supporting the investment and only 637 opposed. Two years after the election, in 1927, the new plant, fully built but not yet operational, was dedicated at an event that featured a program held on the floor of an enormous septic tank, measuring some eighty feet in diameter. Siouxlanders proudly crowded together there, in what would soon become a metal capsule of sewage sludge. Their celebration wouldn't have the panache of the Queen Bee's debut, but unlike toasting a monumental gristmill, the newer event commemorated a project of enduring public value.

Just three years after the new treatment plant opened, more concerns arose about contaminated effluent released into the river by the city and one of the city's most important businesses, the John Morrell livestock slaughterhouse and meatpacking plant. Morrell started its first packing plant in Sioux Falls in 1909, and two years later, the company opened a permanent operation, located along the Big Sioux River downriver from the falls.

The first systematic expression of national interest in protecting water quality was passage of the Federal Water Pollution Control Act in 1948. At the time, more than 3,500 towns and cities across the country were dumping 2.5 billion tons of raw sewage into rivers, streams, lakes and coastal waters every day. Unfortunately, the new authorization lacked muscle and money and failed to curb the problem. Aquatic life continued to perish. Drinking water supplies were compromised. Rivers brimming with chemicals burst into flames. It was estimated that 5,000 U.S. communities lacked adequate wastewater treatment facilities in 1965. Finally, in 1972, a more comprehensive package of policies and practices was adopted, as Congress overrode President Richard Nixon's veto to enact the Clean Water Act. Passage of that historic measure, a huge victory by the rising environmental movement, established water quality guidelines and standards, enforcement provisions and a war chest supplying financial support to communities upgrading their sewage systems, treatment plants and waste management strategies.

The early 1970s revealed new circumstances regarding river pollution in Sioux Falls. Inspectors discovered that nearly forty stormwater outlets within the city, including outlets linked to private businesses, discharged raw or partially treated sewage into the river, and many of those outlets were located upriver from the falls. This so-called blending of waste was common in cities across the country. Steve Metli was outraged, and within two years, the city had addressed the problem.

Another Big Sioux River issue was flood control. A decade before passage of the Clean Water Act and prompted by periodic flooding within Sioux Falls, Big Sioux flows approaching Sioux Falls were redirected away from the city via a diversion dam and a flood prevention bypass channel built by the U.S. Army Corps of Engineers. To further fortify flood prevention, the corps built levees and a gated flood control structure above the mouth of Skunk Creek and modified the Big Sioux channel through the heart of Sioux Falls. These projects reduced the threat of flooding in Sioux Falls and altered the river's flow regime within the city. Skunk Creek, one of the Big Sioux's largest tributaries, would contribute a greater share of water passing over the cascades.

Entering the Big Sioux near the city's zoo, downstream from the diversion dam, Skunk Creek had its own pollution problems. Originating as an outflow at Brandt Lake, near Madison, South Dakota, some thirty-five miles northwest of Sioux Falls, the Skunk drained a sizeable area dominated by agriculture, including intensive grain farming and livestock production. Though a modest waterway, Skunk Creek carried worrisome loads of farm pollution into Sioux Falls.

The Morrell plant continued to stumble through a series of serious water quality violations, highlighted by the 1996 convictions of the plant's general manager and chief engineer for falsifying reports detailing the amount of pollution discharged into the Big Sioux River and knowingly exceeding pollution permits for eight consecutive years. The meatpacking operation, acquired by Virginia-based and Chinese-owned Smithfield Foods in December 1995, agreed to pay a $3 million penalty. The plant's general manager was sentenced to two years in prison and the plant's chief engineer to six months.

Augustana University biology professor Dr. Craig Spencer led a 1996 study of aquatic life in the river that confirmed the river's contaminated condition. "We're spending millions to spruce up Falls Park," Spencer told the *Argus Leader*, "but we have done very little for water quality."

Worrisome conditions continued. Between 2000 and 2019, Smithfield committed at least sixty-one serious pollution violations, and other sources of contamination were also unaddressed. Extensive land use changes were occurring in the Big Sioux watershed, particularly the conversion of native prairie to annually planted crops. This conversion caused significant changes to water quality in the Big Sioux. Farm chemicals, including synthetic fertilizers, pesticides and herbicides, and eroded soils running off grain fields into waterways were not subject to laws and regulations. Tragically, in 2012, the Big Sioux River was rated the thirteenth most polluted river in the United States.

Dana Loeske, founder of a conservation group called Friends of the Big Sioux River, warned that a focused campaign to clean up the river was imperative. "The river is unacceptably polluted," said Loeske in 2017. "It is likely that no river in our state contains water as contaminated as the Big Sioux River."

Leading the charge for cleaner water in the Big Sioux and its watershed was a regional governmental organization, East Dakota Water Development District, based in Brookings, South Dakota. The agency convinced some farmers to move grain crops away from shorelines and to grow riparian buffers instead. This approach reduced chemical and erosion runoff into waterways. Feedlots were relocated away from surface water, and cattle were prevented from entering the river. The improvements to Skunk Creek's water quality were laudable. The City of Sioux Falls contributed to these projects and other improvements.

"Progress has been made," said Travis Entenman, manager of Friends of the Big Sioux River. "But so much more needs to happen. Success will result only when a variety of interests work together. Municipalities and other governmental entities, farmers, industry, businesses, homeowners and groups such as ours must cooperate with one another. We all need to pursue meaningful solutions if lasting benefits and protections can happen." Stronger laws, vigorous enforcement and a more conscientious citizenry will help, said Entenman. "Clean water must become a greater priority," he explained.

Sioux Falls officials estimated that approximately five hundred thousand people visited Falls Park in the year 2024. No one in city hall expected that figure to shrink during ensuing years. Dirty water notwithstanding, the allure of the cascades was undeniable.

It was impossible to know what those Falls Park visitors felt as they stood on a viewing platform inspired by Hazel O'Connor and gazed at the cascades. Did they simply marvel at the scenery and leave the park empty-handed? Or did they desire to understand the ecological and geological character of the place? Did they want to know the human history of the cascades and the instructive lessons of that history? Were any of them inspired to feel reverence for nature and the human relationship to nature?

As a little girl, Hazel O'Connor must have straightened up from her seat in the family's fancy surrey as her father steered them to the cascades. From that vantage, she could more fully admire what she described as the most beautiful sight she'd ever beheld. The water flowing over the falls was crowded by a shoreline littered with gritty industry, and the falls had fallen into slavery. But O'Connor's youthful perspective focused on the roiling, falling water and glossy reddish rock that had survived exploitation and manipulation. That was enough for a little girl.

An older, wiser Hazel O'Connor more closely surveyed the scene, and the invasion of commerce stood out like a black eye and bruises on an infant. O'Connor understood that Sioux Falls suffered when its namesake was diminished and disrespected. She believed her community's values were demonstrated by how the falls fared.

BIBLIOGRAPHY

Books

Ambrose, Stephen. *Undaunted Courage: Meriwether Lewis, Thomas Jefferson, and the Opening of the American West.* Simon and Schuster, 1996.
Armstrong, Moses. *The Early Empire Builders of the Great West.* University Press of the Pacific, 1866.
Bailey, Dana R. *History of Minnehaha County, South Dakota.* Brown and Saenger, 1899.
Blegen, Theodore C. *Minnesota: A History of the State.* University of Minnesota Press, 1975.
Bray, Edmund, and Martha Colman Bray, editors and translators. *Joseph Nicollet on the Plains and Prairies: The Expeditions of 1838–1839.* Minnesota Historical Society Press, 1976.
Bray, Martha Colman, editor. *The Journals of Joseph N. Nicollet: A Scientist on the Mississippi Headwaters. With Notes on Indian Life. 1836-1837.* Minnesota Historical Society Press, 1970.
The Energy to Make Things Better: An Illustrated History of Northern States Power Company. Northern States Power, 1999.
Fanebust, Wayne. *Cavaliers of the Dakota Frontier.* Heritage Books, 2009.
———. *Echoes of November.* Pine Hill Press, 1997.
———. *Where the Sioux River Bends.* Pine Hill Press/Minnehaha County Historical Society, 1985.
Ferris, Jacob. *The States and Territories of the Great West.* Miller, Orton and Mulligan, 1856.

Gottschall, Amos. *Travels from Ocean to Ocean, from the Lakes to the Gulf.* Self-published, 1894.

Gries, John Paul. *Roadside Geology of South Dakota.* Mountain Press, 1996.

History of Southeastern Dakota, Its Settlement and Growth. Western Publishing Company, 1881.

Johnson, Carter, and Dennis H. Knight. *Ecology of Dakota Landscapes.* Yale University Press, 2022.

Kingsbury, George. *History of Dakota Territory.* Vols. 1–4. S.J. Clark Publishing, 1915.

Lauck, Jon K., editor. *Heartland River.* Center for Western Studies, 2022.

Lauck, Jon K., and Patrick Hicks, editors. *City of Hustle: A Sioux Falls Anthology.* Belt Publishing, 2022.

Manfred, Frederick. *Conquering Horse.* University of Nebraska Press, 1983.

Meyer, Herbert H. *Builders of Northern States Power.* Northern States Power, 1972.

Moulton, Gary, editor. *The Definitive Journals of Lewis and Clark.* Bison Books, 2002.

Parker, Donald Dean, editor. *The Recollections of Philander Prescott; Frontiersman of the Old West.* University of Nebraska Press, 1966.

Petersen, William John. *Iowa: The Rivers of Her Valleys.* State Historical Society of Iowa, 1941.

Pielou, E.C. *After the Ice Age.* University of Chicago Press, 1991.

Prescott, Philander. *Autobiography and Reminiscences of Philander Prescott.* Minnesota Historical Society, 2007.

Smith, Charles. *A Comprehensive History of Minnehaha County.* Educator Supply Company, 1949.

Theodore, Christian. *Minnesota: A History of the State.* University of Minnesota Press, 1963.

Thornbury, William D. *Regional Geomorphology of the United States.* John Wiley & Sons, 1965.

Articles, Reports, Documents

Andrews, John. "Pettigrew's Redemption." *South Dakota Magazine* (September/October 2010).

Baldwin, Brewster. *Preliminary Report on the Sioux Quartzite.* South Dakota Geological Survey, 1949.

Bibliography

Captain J. Allen's Expedition. Letter from the Secretary of War. March 20, 1846. Twenty-Ninth Congress, First Session. Doc. No. 168. In compliance with a resolution of the House of Representatives, January 29, 1845.

Flint, Richard Foster. *Pleistocene Geology of Eastern South Dakota.* U.S. Geological Survey, 1955.

Huff, Sanford W., editor. *Navigation of the Missouri.* Annals of Iowa State Historical Society, 1868.

Keyes, Charles Rollin. "Opinions Concerning the Age of the Sioux Quartzite." *Proceedings from the Iowa Academy of Sciences* 2, Annual Issue (1894).

Nicollet, J.N. *Report Intended to Illustrate a Map of the Hydrological Basin of the Upper Mississippi River.* Twenty-Sixth Congress, Second Session, Senate Document 237, 1843.

Nord, David Paul. "The Flour Milling Revolution in America, 1820–1920." *Indiana Magazine of History* (December 2020).

O'Hara, Cleopa, and James E. Todd. *Mineral Resources in South Dakota.* South Dakota Geological Survey, 1902.

Olson, Gary D. "The Queen Bee Legend." *South Dakota History* (Winter 1998).

Pettigrew R.F. "Early Days of Sioux Falls." *Sunshine State Magazine* (November 1925).

Rogers, Dilwyn, J. *Ecology and Natural History of the Sioux Falls Area.* Augustana University, 1971.

Shimek, B. *The Flora of the Sioux Quartzite in Iowa.* Iowa Academy of Sciences, 1896.

Smith, Reed W. "Samuel Medary and the Crisis: Testing the Limits of Press Freedom." PhD diss., Ohio State University, 1995.

Southwick, D.L. *Fluvial Origin of the Lower Proterozoic Sioux Quartzite, Southwestern Minnesota.* 1986.

Southwick, D.L., editor. *Shorter Contributions to the Geology of the Sioux Quartzite (Early Proterozoic), Southwestern Minnesota.* 1984.

Todd, J.E. *A Preliminary Report on the Geology of South Dakota.* South Dakota Geological Survey, 1894.

Tomhave, Dennis. *Geology of Minnehaha County, South Dakota.* South Dakota Geological Survey, 1994.

White, C.A. "A Trip to the Great Red Pipestone Quarry." *American Naturalist* 2, no. 12 (1869).

Witte, Kevin C. "In the Footsteps of the Third Spanish Expedition: James MacKay and John T. Evans' Impact on the Lewis and Clark Expedition." *Great Plains Quarterly* (Spring 2006).

INDEX

A

Albright, Samuel 59, 60
Allen, James 46, 47
American Fur Company 42, 43
Amidon, Joseph 61
Argus Leader newspaper 94, 106, 115, 121, 137, 139
Armstrong, Moses 62

B

Baily, Dana 82
Baldwin, Brewster 24
beaver trapping and business 37, 42, 43
Big Muddy Workshop
  improvement plan for Falls Park, 1990s 140, 141, 142, 143
Big Sioux River 18, 19, 20, 22, 25, 29, 31, 35, 37, 38, 39, 43, 44, 45, 46, 50, 51, 52, 54, 55, 56, 63, 72, 74, 84, 86, 87, 94, 105, 109, 110, 111, 118, 120, 125, 127, 129, 131, 132, 133, 135, 153, 154, 156, 157, 159, 160, 161, 162
  interlobate waterway 25
  pollution 157, 159, 160, 161, 162
  River of the Mahas 38
Blesener, Katie 141, 144
Blood Run village 35, 37, 38
  Omaha tribe 35
Brookings, Wilmot 52, 59, 64, 75
Browning, John 124, 157
Buchanan, James 54
Bulow, W.J. 136
Burlington Northern Sante Fe Railroad (BNSF) 151
Burns, Richard 128
Byllesby, Henry 89, 90, 91, 92, 94, 95, 97
  Consumers Power Company 94
  Northern States Power 94

C

Canadian Shield 20
Cascades Mill 72, 73, 88, 89
 dam 72, 73
Cavelier, René-Robert, Sieur de La Salle 40
Chenandes des Voyageurs 38
Cherapa Place 154, 155
Cimino, Michael 118
Clark, Willam 39, 40
 map and place names for Big Sioux River 39
Clean Water Act of 1972 160
Clovis culture 34, 35
Columbia Fur Company 42
Commander Larabee Elevators, Milling and Storage 116
Confluence (landscape architecture and urban planning firm) 156, 157
Conquering Horse (book by Frederick Manfred) 37, 118
Cooper, Mike 140, 141, 144, 146
Cowman, Tim 22
Crusinberry, V.L. 127

D

Dakota Democrat newspaper 59
Dakota Land Company 52, 53, 54, 55, 56, 57, 58, 59, 60, 63, 68
 townsites 55
Dakota Territory 22, 36, 55, 56, 59, 60, 62, 71
Danz, Jeff 139
Des Moines glacial lobe 25
Drake, E.F. 75

Drake, James 75, 84, 85, 86, 87, 89, 102, 110, 111
Dubuque, Iowa 50, 52

E

earth's crust 20
earth's mantle 20
East Dakota Water Development District 162
Ecker, Peder 139
Edison, Thomas 87, 89, 90
Edmison, Mrs. P.H. 112
Entenman, Travis 162
Environmental Protection Agency
 Seney Island determination 148
Evans, John
 map of middle Missouri River 38, 39

F

Falls Park 113, 114, 115, 117, 118, 121, 122, 123, 128, 129, 130, 131, 132, 133, 135, 137, 138, 139, 140, 141, 144, 146, 148, 149, 150, 151, 152, 153, 154, 156, 157, 161, 162
 Big Muddy plan 140, 141, 142, 143, 144, 145
 Confluence plan 156, 157
 early version 113, 114
Falls Park West 153, 156
Fanebust, Wayne 77, 109, 110, 137
Farrar, Frank 127
Ferris, Jacob 50, 56, 99
Filmore, Millard 54

Fort Dakota 63, 64, 74, 75
Fort Snelling 41, 42, 46, 62
Fort Sod 57
Fremont, John 46, 54
Friends of the Big Sioux River 162

G

General Federation of Women's Clubs 112
geology 17
geomorphology 18
glaciers
    continental 24, 25, 26, 29
    Pleistocene epoch 24
Good Earth State Park 35
Gottschall, Amos 36, 65, 67, 68, 69
    Travels from Ocean to Ocean and from the Lakes to the Gulf 65
granite 20, 21
gristmills 71
    Sioux Falls earliest gristmills 72, 73, 74

H

Hamilton, Alexander 49
Hannus, Adrien 36
Hanson, Gary 146, 148, 150, 151, 152, 156
Hazard, Jeff 152
History Club, the 112, 113, 114, 119, 121, 122
    clubhouse 120
Homestead Act of 1862 49
Hoskins, Bill 151

I

Iowa 21, 22, 23, 52, 61, 62, 121, 130
Izaak Walton League 114, 122, 159, 160

J

Jacobson, Jon 156, 157
Jacobson Plaza 156
James glacial lobe 25, 29
James River lowlands 25
Janklow, Bill 151
Jayne, William 62, 71
Jefferson, Thomas 39
John Morrell meatpacking plant 148, 160, 161
Johnson, Carter 31

K

Kearney, Don 155
Kingsbury, George 36, 51
Kiwanis Park 128, 154
Kneip, Richard 136, 138
Knobe, Rick 131, 132, 133, 136, 138, 139, 140, 156

L

Ladies History Club 105, 106, 112
    becomes History Club 112
LeSueur, Pierre Charles 37, 38
Levitt at the Falls 153
Lloyd, Craig 152, 153

Lloyd Landing 154
Loeske, Dana 162
Lone Rock 79, 80
Louisiana Territory 39, 40
Lower Sioux agency 61
Lundquist, Elof 120
Lundquist, Theresa 119, 120

M

magma 17, 20, 21
Manfred, Frederick 36, 115, 116, 118
Mankato, Minnesota
  execution of Sioux warriors 62
Margulies, Ben 116, 128, 136
Mashek, Carol 125, 139
Masters, Henry 59
McCart, Earl 123, 127, 130, 131, 138, 143
McKinnon, Thomas 106
McQuillen, John 127
Medary, Samuel 53, 54, 56, 57
Medary (townsite) 55
Metli, Steve 131, 132, 140, 142, 144, 146, 149, 152, 153, 161
Millard, Ezra 50, 99
Mills, David 50, 99
Minneapolis, Minnesota 42, 48, 58, 65, 77, 83, 84, 101, 116, 152
Minneapolis Star Tribune 101
Minnehaha Falls (Minneapolis) 58
Minnesota 21, 23, 26, 41, 42, 53, 56, 59
Minnesota River 25, 41, 47, 55, 61
Minnesota Territory 47, 53, 54, 59
Missouri River 19, 38, 39, 40, 43, 45

Monson, Ed 146
Munson, Dave 151, 152

N

Na-he-no-Wenah 41, 42
Nebraska 21, 23
Nelson, Mrs. Joseph 124
Nicollet, Joseph 44, 45, 46
Northern States Power (NSP) 94, 97, 130, 138, 146
  powerhouse 138

O

O'Connor, Barbara 119, 122, 139
O'Connor, Hazel 119, 121, 122, 124, 125, 126, 127, 128, 129, 130, 131, 133, 138, 139, 154, 157, 163
  attending park board meetings 122, 123, 124
  History Club activism 119, 120, 121, 122
  Sioux Steel conflicts 126, 127, 129, 130
O'Connor, Joseph 122, 125
O'Connor, Michael 119
Ogdie, Sam 136, 137
Ogdie, Sam, Jr. 136, 137

P

Pantagraph newspaper (Sioux Falls) 72

Pettigrew, Richard 63, 67, 74, 75, 76, 77, 78, 80, 82, 84, 86, 105, 106, 111, 112, 116, 149
  protests trains near falls 105, 106, 149
  Queen Bee mill 74, 76, 77, 78, 80
  railroad promoter 75
Phillips Avenue 63, 105, 139, 149, 150, 151, 152, 153
Phillips Avenue to the Falls 149, 150, 151, 152
Phillips, Josiah 52, 63, 64, 75, 139
Pillsbury, Charles 83
Pipestone site 24, 35, 47
Pitts Inc. 150
Pleistocene Epoch 24
Potter, Joseph Haydn 47
prairie coteau 19, 25, 26, 29, 43, 45
Prescott, Philander 41, 42, 43, 44, 46, 48, 61, 62, 84
  cascades of the Big Sioux River 43, 44
  killed during Sioux Uprising of 1862 61

Q

Queen Bee mill 74, 76, 77, 78, 80, 81, 82, 83, 84, 90, 91, 101, 103, 110, 116, 128, 130, 138, 143, 155, 160
  engineering specifics 80
  investors 75, 76

R

Raventon, Edward 36

River Improvement Study and Evaluation group (later the River Improvement Society, or RISE) 128, 129, 131, 132, 133, 139
Royster, John 141, 142, 143, 144, 145, 146, 149, 155
Rysdon, Charles 125, 126
Rysdon, Max 126, 127

S

San Antonio Riverwalk 129, 155
Scherschligt, Jeff 154, 155
Schirmer, M.E 131
Seney, George 75, 76, 77, 80, 101, 103, 110, 111
  Queen Bee mill 75, 76
Seney Island 76, 91, 99, 100, 101, 102, 103, 104, 105, 106, 107, 110, 111, 113, 120, 126, 129, 142, 148, 150, 153, 154, 155
  Methodist Church gatherings 103, 104
  threatened by development 102, 104, 105, 106, 107
Seton, Ernest Thompson 37
Sioux City, Iowa 22, 50, 51, 52, 74, 105, 149
Sioux Falls City 56, 57
Sioux Falls Corrugating Company (original name for Sioux Steel company) 125
Sioux Falls Greenway 132, 135, 140, 149, 153, 155, 156
Sioux Falls Light and Power 90, 91, 92, 94
  impacts to the cascades 92

Index

Sioux Falls, South Dakota 23, 43, 59, 60, 61, 62, 63, 64, 65, 76, 77, 78, 80, 83, 84, 85, 86, 87, 89, 90, 94, 99, 101, 102, 103, 104, 105, 107, 109, 110, 111, 112, 113, 114, 115, 116, 117, 118, 119, 120, 121, 122, 123, 126, 127, 128, 130, 131, 132, 133, 135, 136, 137, 140, 141, 148, 151, 152, 155, 156, 159, 160, 161, 162
  early settlement 50, 52, 56, 57, 58, 60, 62, 67
  flood of 1881 78, 79, 80
  geological history 29
  gristmill ambitions 71, 72, 73, 83
  municipal utility 89
  railroad development 73, 74, 75
  water pollution troubles 159, 160, 161, 162
  winter of 1880–81 78
Sioux Indians 36, 37, 41, 42, 47, 48, 56, 57, 68, 69, 118
  early treaties 47, 48
  Sisseton and Wahpeton Sioux 41
Sioux Indian War of 1858 56, 57
Sioux Quartzite 20, 21, 22, 23, 24, 28, 43, 47, 50, 65, 85, 120, 121, 154
  commercial attributes 18, 24, 50, 74
  geological formation 21, 22
  quarries 73, 86, 114, 128
Sioux Steel 125, 126, 127, 129, 130, 153, 154
  impacts on the Big Sioux River 126, 127, 128
Sioux Uprising of 1862 61, 62
Skunk Creek 161, 162

Smith, Charles 99, 107
South Dakota Central Railroad 86
  bridge over the cascades 87, 88, 138
Spencer, Craig 161
Spirit Lake, Iowa massacre of 1857 56
St. Anthony Falls 58, 101
Staples, George 50, 52, 56, 99
Steel District 142, 153, 154, 155, 156
St. Paul, Minnesota 52, 53, 55, 57, 68

T

Todd, James Edward (J.E.) 18, 24, 29

U

Upper Sioux agency 61
Uptown at Falls Park group 152

V

V&O Salvage 136

W

Washington, George
  gristmill 71
Water Pollution Control Act of 1948 160

Webber-Hawthorne
  gristmill 72
Western Town Company 50, 51,
    52, 53, 56, 63, 99
Westinghouse, George, Jr. 89, 90
White, C.A. 22, 74
White, Jack 140
Wilkes, Eliza 111
Wisconsin-era glaciers 25, 26
Woodlawn Cemetery 86

Y

Yankton
  Dakota Territory 62
Yeager, Bert 114

Z

Zip Feeds building 154

ABOUT THE AUTHOR

Courtesy of Zach Carrels.

Peter Carrels is an award-winning environmental writer, educator and activist specializing in descriptions of the human impacts on rivers and watersheds. His work has ranged from detailing and defending the natural characteristics of the James River of the eastern Dakotas to documenting the environmental history of the Missouri River and other major rivers paddled by the Lewis and Clark expedition. He and colleagues at the publication *High Country News* earned a George Polk Award for their investigative reporting about rivers. Carrels's first book, *Uphill Against Water*, has been described as one of the most important books explaining water and river politics in the American West. His partnerships and sponsors have included some of the nation's most influential environmental organizations. Carrels lives in Sioux Falls, South Dakota.